AF327691

GENOMIC *and* NON-GENOMIC EFFECTS *of* ALDOSTERONE

Edited by
Martin Wehling

CRC Press

Boca Raton   Ann Arbor   London   Tokyo

GENOMIC *and* NON-GENOMIC EFFECTS *of* ALDOSTERONE

Library of Congress Cataloging-in-Publication Data

Genomic and non-genomic effects of aldosterone / edited by Martin
  Wehling.
      p.   cm. — (Pharmacology and toxicology)
   Includes bibliographical references and index.
   ISBN 0-8493-8385-4
    1. Aldosterone—Physiological effect. I. Wehling, Martin.
II. Series: Pharmacology & toxicology (Boca Raton, Fla.)
   [DNLM: 1. Aldosterone—pharmacology. 2. Receptors, Aldosterone—metabolism.
3. Receptors, Steroid—metabolism. WK 755 G335 1994]
  QP572.A4G46 1994
  615'.364—dc20
  DNLM/DLC
  for Library of Congress   94-12406
                                                       CIP

PHARMACOLOGY AND TOXICOLOGY: BASIC AND CLINICAL ASPECTS

Mannfred A. Hollinger, Series Editor
University of California, Davis

Published Titles
Angiotensin II Receptors: Molecular Biology, Biochemistry, and Pharmacology, Robert J. Ruffolo, Jr.
Basis of Toxicity Testing, Donald J. Ecobichon
Beneficial and Toxic Effects of Aspirin, Susan E. Feinman
Biopharmaceutics of Ocular Drug Delivery, Peter Edman
Direct Allosteric Control of Glutamate Receptors, M. Palfreyman, I. Reynolds, and P. Skolnik
Genomic and Non-Genomic Effects of Aldosterone, Martin Wehling
Human Drug Metabolism from Molecular Biology to Man, Elizabeth Jeffreys
Inflammatory Cells and Mediators in Bronchial Asthma, Dvendra K. Agrawal and Robert G. Townley
In Vitro *Methods of Toxicology*, Ronald R. Watson
Peroxisome Proliferators: Unique Inducers of Drug Metabolizing Enzymes, David E. Moody
Pharmacology of the Skin, Hasan Mukhtar
Platelet Activating Factor Receptor: Signal Mechanisms and Molecular Biology, Shivendra D. Shukla
Preclinical and Clinical Modulation of Anticancer Drugs, Kenneth D. Tew, Peter J. Houghton, and Janet A. Houghton

Forthcoming Titles
Alternative Methodologies for the Safety Evaluation of Chemicals in the Cosmetic Industry, Nicola Loprieno
Animal Models of Mucosal Inflammation, Timothy S. Gaginella
Brain Mechanisms and Psychotropic Drugs, A. Baskys and G. Remington
Drug Delivery Systems, V. V. Ranade
Human Growth Hormone Pharmacology: Basic and Clinical Aspects, Kathleen T. Shiverick and Arlan Rosenbloom
Immunopharmaceuticals, Edward S. Kimball
Neural Control of Airways, Peter J. Barnes
Pharmacology Effects of Ethanol on the Nervous System, Richard A. Deitrich
Pharmacology in Exercise and Sports, Satu M. Somani
Pharmacology of Intestinal Secretion, Timothy S. Gaginella
Phospholipase A2 in Clinical Inflammation: Endogenous Regulation and Pathophysiological Actions, Keith B. Glaser and Peter Vadas
Placental Pharmacology, B. B. Rama Sastry

PREFACE

Aldosterone is the main mineralocorticoid in humans. As its major physiological function, it regulates the electrolyte and water balances by renal effects. Being a typical steroid hormone, its mechanism of action was explained by the common theory of *genomic* steroid effects. This theory was developed some thirty years ago and relates steroid action to receptor-mediated hormone effects on nuclear DNA and subsequent synthesis of proteins. These *genomic* steroid effects are characterized by a delayed onset and, thus, clearly distinct from instant, *non-genomic* effects such as catecholamine action on transmembrane ion fluxes, and the related physiological responses including positive cardiac inotropy. Over the past few years increasing evidence has accumulated for rapid steroid effects which are not compatible with this genomic theory of steroid action and have been neglected, probably reflecting the lack of a sufficient theoretical background. Such rapid effects were also found for aldosterone and extensively studied with regard to their physiological and pharmacological properties. At present, aldosterone appears to be the most prominent example of a steroid hormone for which both the *genomic* and *non-genomic* pathways of action have been defined and characterized.

The aim of this book is to summarize known features of both the more traditional genomic aldosterone effects, which have been known for more than thirty years, and those modern aspects of non-genomic aldosterone effects which have come to our attention just a few years ago. International specialists contribute their updates on these issues, including chapters on traditional genomic mineralocorticoid physiology, mineralocorticoid receptor theory, and rapid non-genomic steroid effects in the brain, kidneys, smooth muscle cells, and lymphocytes.

The authors hope that the new dual model for steroid action presented here may stimulate the efforts of the scientific community to further explore the relevance of these rapid steroid effects for human physiology and pathology.

M. Wehling
Munich, May 4, 1994

THE EDITOR

Martin Wehling, M.D., is senior lecturer (Privatdozent), and head of the division of Clinical Pharmacology, Medizinische Klinik, Klinikum Innenstadt, University of Munich, Germany.

Dr. Wehling graduated, and obtained his M.D. degree, from Kiel University in 1982 as a recipient of a scholarship by the Studienstiftung des Deutschen Volkes. In 1992, he received the degree of Habilitation and became Privatdozent (senior lecturer) at Munich University. He received a "Heisenberg"-scholarship by the Deutsche Forschungsgemeinschaft in 1993, and spent a sabbatical at the Baker Medical Research Institute in Melbourne with John Funder. In 1994, he was elected head of the Klinische Forschergruppe, "Clinical Pharmacology," funded by the Deutsche Forschungsgemeinschaft, and head of the division of Clinical Pharmacology at the Medizinische Klinik, Klinikum Innenstadt, University of Munich.

He holds board certificates as an internist, cardiologist, and clinical pharmacologist, and is a member of the German societies of cardiology, pharmacology, endocrinology, and internal medicine.

Dr. Wehling has published over 150 research papers and 2 books. His current major interest is focused on further work related to rapid, nongenomic steroid effects.

CONTRIBUTORS

Catherine Barlet-Bas, Ph.D.
Laboratoire de Physiologie Cellulaire
Collège de France
Paris, France

Alain Doucet, Ph.D.
Laboratoire de Physiologie Cellulaire
Collège de France
Paris, France

John W. Funder, M.D., Ph.D.
Baker Medical Research Institute
Melbourne, Australia

Bruce S. McEwen, Ph.D.
Laboratory of Neuroendocrinology
Rockefeller University
New York, New York

Hans Oberleithner, M.D.
Physiologisches Insitut
Universität Würzburg
Würzburg, Germany

Miles Orchinik, Ph.D.
Laboratory of Neuroendocrinology
Rockefeller University
New York, New York

Albrecht Schwab, M.D.
Physiologisches Insitut
Universität Würzburg
Würzburg, Germany

Martin Wehling, M.D.
Division of Clinical Pharmacology
University of Munich
Munich, Germany

CONTENTS

LAST BUT NOT LEAST
IN THE GUISE OF AN INTRODUCTION

Why such a title? When Dr. Wehling invited me to write introductory remarks for this series of scientific contributions on "heterodox" effects of corticosteroids, he obligingly let me have access to the chapters he and his colleagues had written. I thought that in this way the risk of making irrelevant comments would thus be decreased…and it meant that my contribution, small though it be, landed on the Editor's desk last.

What about least then? Rapid steroid effects have a descriptive connotation: the period of time elapsing between exposure to steroid and the physiological consequences is deemed too short for "the dogma" — to quote Dr. Wehling — to apply. Indeed, inasmuch as steroids act by reflecting an interaction of the steroid-receptor complex with genomic DNA, a significant latency period is the rule. In such a context the notion of "rapid" is relative, as we are cogently reminded of by Drs. Orchinik and McEwen.

In the case of cortisol, the main glucocorticoid in the human, "the dogma" does not apply to the quasi-immediate inhibition of release of corticotropin as the concentration of cortisol in the blood stream is abruptly increased (cf Fig. 4.2, in Myles and Daly, 1974). The converse is also observed, in that secretory activity of these anterior pituitary cells reappears or increases as soon as extracellular cortisol concentrations fall below threshold — provided that the duration of exposure of the corticotrophs to high levels of cortisol is kept short.

That this case of cortisol and corticotropin prompted a search for other, equally rapid steroid effects is very understandable; yet, to date compelling data are few and far between. When it comes to aldosterone, the main mineralocorticoid in the human, in terms of time course we are still a long way from having an equivalent *in vivo* response to that of corticotrophs to cortisol.

Our readiness to extrapolate from simpler, more manageable preparations may in fact lead us away from what is ultimately relevant to physiology. At the risk of being dubbed a Doubting Thomas, I would cite the caveats voiced by Drs. Schwab and Oberleithner on the extrapolation from intracellular pH shifts observed shortly after exposing cultured dog renal tubule cells to aldosterone, and the effects of the hormone on the kidney *in vivo*. As they stress in their chapter, the gap in time between cell alkalinization — which they have documented as an early event — and changes in urine electrolyte excretion which occur long afterwards (cf Fig 1 in Crabbé, 1992) — leaves us currently in the dark in terms of the intermediate step(s).

Last but not least, the authors represented in these pages have done us a double service. Not only have they outlined the present state of knowledge of rapid steroid hormone effects, but also they have acknowledged the currently fragmentary, albeit compelling nature of the data which necessitate an additional model of steroid action, over and above the "dogma" of transcriptional regulation.

J. Crabbé

Brussels, June 1994

REFERENCES

Crabbé, J., Aldosterone action and the cost of flame photometry. *Mol. Cell. Endocrinol.,* 90: C11–C13, 1992.

Myles, A. B. and J. R. Daly, "Corticosteroid and ACTH Treatment," Edward Arnold, London, 1974, 216 pp.

To my mother and father

Chapter 1

CLASSICAL MINERALOCORTICOID RECEPTORS AND SPECIFICITY-CONFERRING MECHANISMS

John W. Funder

TABLE OF CONTENTS

I. BACKGROUND

At the outset, the apparent redundancy and thus possible ambiguity in the title of this chapter needs to be addressed, and its implications briefly explored. Normally, we think of receptors as molecules which distinguish signals in two ways. First, most receptors discriminate between potential signals on the basis of their relative affinities, the likelihood of the receptor being occupied by a particular ligand at a particular concentration, the property of specificity. Second, for most receptors some ligands are capable of activating the receptor, so that the signal is propagated, and others, however well they may bind, are not. The first class are agonist compounds, and the second are antagonists. Classical mineralocorticoid receptors conform with this description to a considerable degree, both in that they show clear selectivity (for example, they have a much higher affinity for aldosterone than estradiol-17β) and in that aldosterone antagonists have been widely and successfully used for many years in the clinic.

In physiological terms mineralocorticoid receptors depart from this norm in two important ways. First, mineralocorticoid receptors appear to have equivalent affinity for the physiological glucocorticoids (cortisol, or corticosterone in rats and mice) as for aldosterone. This is true for mineralocorticoid receptors in man,[1,2] rat,[3] and guinea-pig,[4] and represents a particular challenge in terms of aldosterone occupancy of such receptors, in that free concentrations of circulating glucocorticoids are 1 to 2 orders of magnitude higher than those of aldosterone. Secondly, whereas in epithelial tissues mineralocorticoid receptors appear to be equivalently activated by aldosterone and physiological glucocorticoids in terms of the benchmark mineralocorticoid effect (transepithelial Na^+ transport), in nonepithelial tissues this is commonly not the case. If aldosterone has effects clearly mediated via such receptors, corticosterone antagonizes such effects;[5] if corticosterone has an agonist effect, aldosterone is an antagonist.[6]

In the discussion to follow, these two enigmas will occur as a leitmotiv. One is the means whereby a non-selective receptor can mediate physiological responses in aldosterone target tissue, which we are some way towards understanding. The other is the physiological roles for such receptors in nonepithelial tissues, in which they do not appear protected by the specificity-conferring mechanisms found in epithelial tissues. On the contrary, under normal circumstances these receptors are predominantly occupied by cortisol/corticosterone, and thus operate as high affinity glucocorticoid receptors.

II. INITIAL STUDIES

The first reports of saturable binding of [³H]aldosterone in mineralocorticoid target tissues in kidney cytosol preparations[7] and parotid minces[8] from adrenalectomized rats were published in 1972. Given the now notorious instability of mineralocorticoid receptors in broken cell preparations without molybdate ions, the cytosol findings reflect perhaps more good luck than good management. In any event, both studies showed [³H]aldosterone to bind to two classes of relatively high affinity sites, which were subsequently[9] termed Type I sites (higher affinity) and Type II sites (lower affinity). These studies on kidney slices from adrenalectomized rats showed [³H]aldosterone binding to be displaced *in vitro* by corticosterone with approximately one tenth the potency of nonradioactive aldosterone.

Even with an apparent tenfold difference in affinity, the higher circulating levels of glucocorticoid were recognized as a serious potential problem to physiological occupancy of the receptors by aldosterone. The same series of studies[9] addressed this question by parallel experiments on the potency of corticosterone and aldosterone to compete for renal [³H]aldosterone binding *in vivo,* in adrenalectomized rats infused with radioactive and competing steroids before sacrifice. Under such circumstances corticosterone appeared a significantly weaker competitor than *in vitro;* these findings were interpreted as evidence that the much higher plasma binding of corticosterone than aldosterone was a key determinant in allowing aldosterone access to classical mineralocorticoid receptors. In subsequent studies, the Type I receptors (affinity for aldosterone 1 to 2 n*M*) were confirmed as physiological mineralocorticoid receptors,[10] and the Type II receptors (affinity for aldosterone ~80 n*M*) as classical, dexamethasone-binding glucocorticoid receptors.[11] Such classical glucocorticoid receptors have moderately high (~3 n*M*) affinity for dexamethasone, and rather lower (10 to 20 n*M*) affinity for the physiological glucocorticoids cortisol and corticosterone. They appear essentially ubiquitous and have been found in all tissues and cells examined to date, except those of the intact pituitary intermediate lobe.

Early studies on the *in vivo* uptake and retention of glucocorticoids by the rat brain used [³H]corticosterone as tracer.[12] In subsequent studies from the same laboratory with [³H]dexamethasone as tracer,[13] it became clear that the anatomical patterns of labeling differed very substantially for the two tracers. The synthetic glucocorticoid dexamethasone was concentrated, after *in vivo* injection, in the pituitary and hypothalamus with relatively modest uptake into the hippocampus. When corticosterone was injected, the reverse was the case — relatively slight uptake into the hypothalamus and pituitary, but much stronger labeling of the septum and hippocampus. This discrepancy was explored in *in vitro* studies on the hippocampus, which distinguished two classes of binding sites: classical glucocorticoid receptors, with higher affinity for dexamethasone than for corticosterone, and corticosterone-preferring sites, which had higher affinity for corticosterone than for dexamethasone, *in vitro* as *in vivo.* The first studies showing that such corticosterone-preferring sites were receptors capable of transducing a signal were published in 1981. In adrenalectomized rats, hippocampal levels of synapsin, an 80K phosphoprotein, fall; they are restored by administration of corticosterone — but not by equivalent amounts of dexamethasone — consistent with an action via a physiological glucocorticoid receptor other than the classical, dexamethasone binding receptor.[13]

III. MINERALOCORTICOID RECEPTORS AND CORTICOSTERONE-PREFERRING SITES

It was simultaneously reported by several laboratories that hippocampal corticosterone-preferring sites had indistinguishable affinity for corticosterone and aldosterone.[4,14,15] In addition, and very importantly, the apparent aldosterone-selectivity of mineralocorticoid receptors in renal molybdate-containing cytosol preparations was shown to represent contamination by extravascular transcortin.[3] When receptors were separated from transcortin by adsorption onto hydroxylapatite, the hierarchy of affinity for a range of steroids (aldosterone, corticosterone, deoxycorticosterone, cortisol, dexamethasone, 9αfluorocortisol) was shown to be equivalent in hippocampus and kidney. On this basis, it was proposed that binding reflected the presence of Type I receptors in both tissues, with those in the kidney representing physiological mineralocorticoid receptors, and those in the hippocampus a second, higher-affinity (1 to 2 nM vs. 10 to 20 nM) class of physiological glucocorticoid receptors.

In vitro, cytosol preparations from rat kidney and hippocampus bind [3H]aldosterone and [3H]corticosterone with equal affinity when binding to transcortin is prevented either by its being removed[3] or blocked.[16] *In vivo* studies, in contrast, show a clear distinction between the classic mineralocorticoid target tissues (kidney, colon), and nonepithelial tissues such as hippocampus.[17,18] Rats do not synthesize transcortin between 4 and 14 days after birth, so that during this period plasma binding of corticosterone falls to modest levels. One-day adrenalectomized, ten-day old rats were injected with either [3H]aldosterone or [3H]corticosterone, plus excess RU28362 to exclude tracer from classical glucocorticoid receptors, and sacrificed 15 min later. Hippocampal binding of the two tracers was indistinguishable; that of [3H]corticosterone in kidney or colon, however, was ≥ tenfold lower than that of [3H]aldosterone, consistent with the operation of a prereceptor specificity-conferring mechanism excluding glucocorticoids *in vivo* from mineralocorticoid receptors in physiological aldosterone target tissues.

The classical mineralocorticoid receptor was first cloned from a human kidney library in 1987,[2] and shown to be a member of what is now known to be a steroid/thyroid/retinoid/orphan receptor gene superfamily. The human mineralocorticoid receptor is 984 amino acids in length, the largest of the vertebrate receptor family cloned to date. The rat homologue was subsequently cloned from a hippocampal library,[19] testimony again to the commonality of the Type I receptor in both tissues. The mineralocorticoid receptor is clearly a member of a sub-family of steroid receptors which also includes classical glucocorticoid receptors,[20] progesterone receptors,[21] and androgen receptors,[22] which all share ≥50% amino acid identity in their ligand-binding domain and ≥90% identity in the DNA-binding domain. In the initial report on the cloned receptor, high levels of expression were noted in the hippocampus. Similarly, although expressed human mineralocorticoid receptors were shown to bind [3H]aldosterone with high affinity, they were also shown to have equivalent affinity for aldosterone, corticosterone and cortisol.

IV. SPECIFICITY CONFERRING MECHANISMS — 11βHYDROXYSTEROID DEHYDROGENASE

In summary, in 1988 the classical mineralocorticoid receptor had been cloned and sequenced. It had also been shown to be expressed at high levels in nonepithelial as well as epithelial tissue, and to have the same intrinsic high affinity for physiological mineralocorticoids and glucocorticoids. *In vivo,* however, such receptors in kidney and colon were clearly aldosterone selective. One crucial question, then, is how epithelial tissues exclude physiological glucocorticoids from binding to mineralocorticoid receptors.

The answer, or at least part of the answer, to this question was prompted by clinical studies on patients with the very rare syndrome labeled apparent mineralocorticoid excess.[23-25] Such patients present with markedly elevated blood pressure and acute sensitivity to salt, consistent with autonomous overproduction of mineralocorticoids; however renin and aldosterone are suppressed, and cortisol levels normal. In contrast, this is not the case for the patient's urinary steroid profiles, which showed a consistent spectrum of abnormal metabolites, including very elevated ratios of 5α to 5β reduced compounds, and of 11-hydroxy to 11-keto species. A forme-fruste of the disorder could be reproduced by licorice administration, shown to impair the activity of 11βhydroxysteroid dehydrogenase.[26] Taken together, these observations were interpreted[26] as suggesting low 11βhydroxysteroid dehydrogenase activity (either congenitally, as in apparent mineralocorticoid excess, or following licorice ingestion) caused abnormally high intrarenal cortisol levels, which overcame the normal specificity-conferring mechanisms excluding cortisol from mineralocorticoid receptors in epithelia, and thus caused sodium retention and high blood pressure.

What appears to be the case is that the operation of the enzyme 11βhydroxysteroid dehydrogenase is itself the normal specificity-conferring mechanism in aldosterone target-tissues, at least as far as excluding glucocorticoid hormones is concerned.[27,28] When 1-d adrenalectomized, 10-d old rats are pretreated with carbenoxolone (the hemisuccinate of glycyrrhetinic acid, the active principle of licorice) there is no change in the hippocampal uptake or retention of injected [^{3}H]aldosterone or [^{3}H]corticosterone, compared with control, non-pretreated rats. In the epithelial tissues, however, carbenoxolone produces a dramatic effect, increasing [^{3}H]corticosterone binding ~tenfold, to levels equal to (kidney) or approaching (colon) those of [^{3}H]aldosterone.[27] 11βhydroxysteroid dehydrogenase activity, then, appears to be the mechanism excluding corticosterone from epithelial mineralocorticoid receptors in the rat. A similar interpretation was drawn from parallel studies on kidney slices, in which enzyme activity was blocked by incubation with glycyrrhizic acid.[28]

In vitro, minces or homogenates of kidney and colon were shown to be very active in converting 11-hydroxy to 11-keto steroids, and equivalent preparations from hippocampus were much less active. Aldosterone is not a substrate for 11βhydroxysteroid dehydrogenase, presumably reflecting the cyclization of the hydroxyl at C11 with the unique and very reactive aldehyde group at C18, which yields the 11,18 hemiketal configuration. Also shown in these original studies[27] is that the process of 11-dehydrogenation substantially lowers affinity for mineralocorticoid receptors, so that 11 dehydrocorticosterone has only 0.3% the affinity of corticosterone. Of considerable physiological importance, as will be seen later, is a similar loss of affinity for classical glucocorticoid receptors.

Some time before the studies proposing 11βhydroxysteroid dehydrogenase as the mechanism of aldosterone selectivity in mineralocorticoid target tissues, the enzyme had been purified from rat liver[29] and subsequently cloned.[30] Antisera produced to the purified liver enzyme showed high levels of activity in liver, testis, lung, and kidney. In the kidney, however, activity appeared confined to the proximal convoluted tubules rather than the distal tubular elements known to be the site of action of aldosterone,[31] and led to the suggestion that a second species of 11βhydroxysteroid dehydrogenase was responsible for protecting physiological mineralocorticoid receptors in aldosterone target tissue.[32]

V. SPECIFICITY CONFERRING MECHANISMS — HORMONE RESPONSE ELEMENTS

As noted earlier, conversion of corticosterone to 11-dehydrocorticosterone or cortisol to cortisone not only very markedly lowers affinity for mineralocorticoid receptors, but also for classical glucocorticoid receptors. Secondly, in a series of studies on the index patient with apparent mineralocorticoid excess, cortisol was shown to be more potent than aldosterone in terms of sodium retention and blood pressure elevation.[33] The latter is difficult to reconcile with an effect of cortisol uniquely through unprotected mineralocorticoid receptors in such patients, but would be consistent with an effect via both mineralocorticoid and glucocorticoid receptors, for which aldosterone has much lower affinity.

To test this possibility, patients with pseudohypoaldosteronism were given carbenoxolone to block renal 11βhydroxysteroid dehydrogenase. If the effects of cortisol were uniquely via mineralocorticoid receptors, carbenoxolone would be predicted to be without effect, as such patients have no mineralocorticoid receptors on ligand-binding studies.[34] On the other hand, if the defect in apparent mineralocorticoid excess reflected inappropriate cortisol access to mineralocorticoid and glucocorticoid receptors, both normally protected by 11βhydroxysteroid dehydrogenase, then carbenoxolone administration might have some effect. The second of these possibilities proved to be the case: over two weeks of carbenoxolone administration to four patients with pseudohypoaldosteronism, all proven receptor negative, urinary Na$^+$ to creatinine ratios fell, urinary K$^+$/Na$^+$ rose, and plasma bicarbonate promptly rose.[35] These results thus appear consistent with a mineralocorticoid action of cortisol via classical glucocorticoid receptors, in turn suggesting that the genomic response elements involved in initiating the physiological response to aldosterone may be essentially indiscriminantly activated by mineralocorticoid or glucocorticoid receptors.

This interpretation was strengthened by two additional series of studies, one *in vivo*[35] and the other *in vitro*.[36] Adrenalectomized rats given RU28362, a highly specific classical glucocorticoid receptor agonist, respond by modest but significant urinary electrolyte change. When the animals are pretreated with carbenoxolone, the urinary response to RU28362 is indistinguishable from that to aldosterone. Corticosterone given to carbenoxolone pretreated rats similarly affects the urinary electrolyte ratio; this effect is not blocked by administration of RU28318 (a potent Type I receptor antagonist) or RU38486 (a potent Type II, classical glucocorticoid receptor antagonist) administered alone, but is completely blocked when RU28318 and RU38486 are administered together. Secondly, rat or rabbit isolated cortical collecting tubule preparations, grown to confluence on a grid and forming a seal between two chambers, pump Na^+ in one direction and K^+ in the other, thus establishing a potential difference between the two chambers that can be measured as a short-circuit current. In such preparations, aldosterone (5 nM), dexamethasone (50 nM) or RU28362 (50 nM) have indistinguishable effects on ion flux and short-circuit current, all causing a ~threefold increase in each parameter.[36]

In summary, there is ample evidence *in vivo* and *in vitro* that the classical effect of aldosterone on urinary electrolyte flux in the kidney may be produced by activation of either mineralocorticoid or glucocorticoid receptors. Such findings suggest that the response element involved may be a canonical imperfect palindrome pentadecamer sequence, of the type GGTACAnnnTGTTCT, which has been shown to be similarly responsive to activated mineralocorticoid receptors, glucocorticoid receptors, androgen receptors and progesterone receptors. There may be more mineralocorticoid receptor selective response elements — perhaps, for example, reflecting subtle differences from the pentadecamer consensus hormone response element described above — which is suggested by recent findings in NRK-1 cells, a cell line derived from rabbit kidney and containing both mineralocorticoid and glucocorticoid receptors. In this cell line, aldosterone is clearly much more potent than dexamethasone in elevating levels of mRNA for the α_1 subunit of Na^+/K^+ ATPase.[37]

More extensively characterized to date — and perhaps indeed an explanation of the Na^+/K^+ ATPase α_1 subunit findings — are the differences between mineralocorticoid and glucocorticoid receptors at the so-called composite response elements, sequences which bind and can respond to both steroid receptors and the transcription factor AP-1, commonly a cFos/cJun heterodimer. One such composite response element, termed plfG, binds steroid receptors with about one tenth the affinity of canonical hormone response elements. What is clear, is that activation of mineralocorticoid and glucocorticoid receptors produces a very different pattern of response to AP-1 activation.[38] AP-1 induced transcription is blocked by glucocorticoid receptor activation via an interaction site in the N-terminal domain, an effect which is not obtained with mineralocorticoid receptor activation. Concurrent mineralocorticoid receptor activation appears to blunt the glucocorticoid receptor mediated effect. Whether this represents competition for the steroid binding site on the composite response element,[39] or an effect due to the formation of mineralocorticoid receptor-glucocorticoid receptor heterodimers, awaits exploration. The existence of such composite response elements is not confined to corticosteroids; a similar element distinguishing between androgen and glucocorticoid receptors has also been described.[40]

VI. TYPE I RECEPTORS IN THE CENTRAL NERVOUS SYSTEM

Although in epithelial tissues activated glucocorticoid receptors appear able to mimic mineralocorticoid receptor activation, this is clearly not the case for nonepithelial tissues. As previously mentioned, dexamethasone given to adrenalectomized rats does not restore hippocampal synapsin,[13] whereas corticosterone does; similarly, the effects of corticosterone on hippocampal $5HT_1$ receptors are not mimicked by dexamethasone.[6] When aldosterone is infused intracerebroventricularly to uninephrectomized rats drinking 0.9% NaCl solution, their blood pressure rises, an effect which is clearly via Type I receptor occupancy, and is not mimicked by infusion of Type II receptor selective glucocorticoid agonists.[5] Such divergent effects of mineralocorticoid receptor and glucocorticoid receptor activation are thus very difficult to reconcile with actions via a canonical, nondiscriminatory response element in the hippocampal or circumventricular cells. They do not, of course, rule out other actions equally responsive to mineralocorticoid or glucocorticoid receptor occupancy, as it is entirely possible that a variety of response elements — some operationally selective, others promiscuous — may exist in the brain.

The second thing that distinguishes epithelial and nonepithelial aldosterone receptors is that whereas in epithelial tissues (or in cell transfection systems) mineralocorticoid receptors can be equally well activated by aldosterone or corticosterone/cortisol (providing that the latter can access the receptors), this is not the case for Type I receptors in the brain. In the studies characterizing the serotonin $5HT_{1a}$ receptor response to adrenalectomy,[6] corticosterone, but not aldosterone, restored receptor levels to those seen in intact animals; when the two steroids were coadministered, however, aldosterone antagonized the effect of corticosterone. Similarly, as noted above, aldosterone raises the blood pressure of uninephrectomized salt-loaded rats when infused intracerebroventricularly, an effect not mimicked by corticosterone infusion. When the steroids are confused, however, corticosterone antagonizes the hypertensive effect of aldosterone at dose levels consistent with an effect via unprotected Type I receptors.

Again, the demonstration of some opposing actions of corticosteroids via mineralocorticoid receptors in the brain is not evidence that all such receptors see corticosterone/cortisol as agonist, and aldosterone as antagonist, or vice versa. There appears to be no difference in translation products of the mineralocorticoid receptor gene between kidney and hippocampus, although differences in initiation sites have been described, consistent with tissue-specific differences in control of receptor gene expression.[41] The only distinction between mineralocorticoid receptors in kidney and hippocampus reported to date is by fast protein liquid chromatography, where hippocampal receptors loaded with [³H]aldosterone or [³H]corticosterone reproducibly ran 1 to 2 fractions ahead of renal receptors, consistent with the binding of a ~15K molecular weight accessory protein in the hippocampal cytosols.[42]

VII. SPECIFICITY CONFERRING MECHANISMS — 11βHYDROXYSTEROID DEHYDROGENASE

As noted above, the studies identifying 11βhydroxysteroid dehydrogenase as a crucial specificity conferring mechanism for otherwise nonselective mineralocorticoid receptors were prompted by clinical observations on the syndrome of apparent mineralocorticoid excess. In the index case, and in most patients subsequently studied, the defect appears to be unidirectional (conversion of cortisol to cortisone, i.e., dehydrogenation), with reductase activity being indistinguishable from normal. Consistent with this being a derangement of enzyme function rather than a congenital absence of expression, in five patients with the syndrome, the sequence of 11βhydroxysteroid dehydrogenase appeared normal.[43] In addition, a second form of apparent mineralocorticoid excess has been described, termed AME Type 2 to distinguish it from that originally described, in which both dehydrogenation and reduction appear deficient, so that while clearance rates of both cortisol and cortisone are lowered, the ratio of urinary metabolites is not grossly different from normal, in contrast with AME Type 1.[44]

What is of particular interest is that licorice administration produces an increase in urinary cortisol to cortisone metabolite ratio, reminiscent of AME Type 1, whereas carbenoxolone, the hemisuccinate of glycyrrhetinic acid, appears to affect both dehydrogenation and reduction, and produces urinary metabolite ratios similar to AME Type 2, i.e., relatively unchanged. In *in vivo* studies on rats, these agents have been shown to affect other enzymes — e.g., 5β reductase — activity of which is also grossly altered in the clinical syndrome of apparent mineralocorticoid excess, prompting a search for endogenous agents in AME which might be responsible for the spectrum of licorice/carbenoxolone effects. To date, the most promising candidate molecules appear to be amidated bile acids, which have been shown to affect 11βhydroxysteroid dehydrogenase activity at submicromolar concentrations.[45]

While the etiology of the clinical syndromes of AME Type 1 and 2 remain to be resolved, there has been considerable progress in terms of the physiological roles of 11βhydroxysteroid dehydrogenase. The species purified[29] and cloned[30] from rat liver has been shown to have a relatively high (micromolar) K_m for cortisol and corticosterone, and is now widely believed to modulate glucocorticoid occupancy of glucocorticoid receptors.[46] Although when expressed in a variety of cells, including a toad bladder cell line,[47] the predominant activity is in the reductase direction, the enzyme clearly catalyzes conversion in either direction with the balance depending on a variety of factors which have not been systematically addressed to date.

The species of 11βhydroxysteroid dehydrogenase responsible for exclusion of glucocorticoids from both mineralocorticoid and glucocorticoid receptors in aldosterone target tissues is now recognized on a number of criteria to be distinct from the cloned species, and is thus termed 11βhydroxysteroid dehydrogenase 2. It has a much lower K^m (~15 to 40 nM) for corticosterone/cortisol,[48] appears to operate uniquely as a dehydrogenase,[48] and is NAD rather than NADP-dependent.[48,49] Although there are reports of purification[50] and sequencing[51] of NAD-preferring species in human and sheep, it is unclear whether this activity is equivalent to the species demonstrated in renal tubular preparations.[48,49]

Whereas the activity of 11βhydroxysteroid dehydrogenase 2 may be sufficient to exclude cortisol/corticosterone from mineralocorticoid receptors in aldosterone target tissues, it is unlikely to be of significant importance in excluding steroids which are not hydroxylated at C11, such as progesterone and deoxycorticosterone, both of which have high affinity for mineralocorticoid receptors, and which in the case of progesterone, circulate at far higher levels than those of aldosterone, in pregnancy not dissimilar to those of cortisol. Whereas classically progesterone has been thought to have ~15% of the affinity of aldosterone for mineralocorticoid receptors,[2] more recently a claim has been made that it has 100× the affinity of aldosterone.[52] The discrepancy between these two values may reflect to some extent the difficulties in working *in vitro* with progesterone. In very careful studies, using siliconized glassware and tracking steroid recoveries, we have found that rat mineralocorticoid receptors have three times higher affinity for progesterone as for aldosterone; in the guinea-pig, the affinities are equal.[68] Given this at least equivalent affinity, and the very high circulating levels of progesterone in the luteal phase and particularly in pregnancy, there may exist specificity-conferring mechanisms in 11βhydroxysteroid dehydrogenase to exclude progesterone (and also possibly deoxycorticosterone) from renal mineralocorticoid receptors. Given the patterns of steroid metabolism in the rat kidney,[53] it has been proposed that 20β reduction may be the process involved.[54] This hypothesis is currently under active examination.

VIII. MINERALOCORTICOID RECEPTORS 1994

Of all the members of the steroid/thyroid/retinoid receptor superfamily, classical mineralocorticoid receptors have clearly received the least attention at the molecular and cellular level. It is difficult to determine the extent to which this reflects their lability, the perceived narrowness of mineralocorticoid action (to promote unidirectional transepithelial flux), or the ambiguities between aldosterone-activated effects in epithelia and putatively glucocorticoid-activated effects in non-epithelial tissues. In addition, effective mineralocorticoid receptor agonists have been available for around three decades, considerably in advance of other areas, so that investigation driven by the possibility of defining antagonists is less likely to be the case. Whatever the cause, there have been relatively few studies addressed to exploring mineralocorticoid receptor action in either epithelial or nonepithelial tissues over the past 5 years. Three such studies, on the guinea-pig, on mineralocorticoid receptor-glucocorticoid receptor heterodimerization, and on mineralocorticoid effects on the heart will be described in some detail, as illustrative of areas of activity in the field of mineralocorticoid effector mechanisms.

First, there are a number of lines of evidence that the guinea-pig has considerable differences in its pituitary-adrenal axis from other species. In brief, it has a Ala24Pro substitution at residue 24 in ACTH,[55] caused by a point mutation (C to G) at position 442 in the POMC sequence. This difference in the hitherto invariant steroidogenic 1 to 24 region of ACTH produces a superagonist species, which circulates at unremarkable levels but sustains a very high cortisol secretion rate from the adrenals. Circulating cortisol is high, and free cortisol even higher, reflecting the relatively low affinity and capacity of guinea-pig transcortin.[56] Classical glucocorticoid receptors have very low affinity, approximately one-twentieth that in the mouse,[57] reflecting a number of sequence differences with other glucocorticoid receptors, most notably the substitution of tryptophan for an otherwise invariant cysteine in the ligand-binding domain.[58] Remarkably, the mineralocorticoid receptor in the guinea-pig is indistinguishable from that in the rat in terms of affinity for aldosterone, steroid specificity, and tissue distribution.[4]

The finding of a pristine mineralocorticoid receptor in the guinea-pig, with very high (1 to 2 nM) affinity for both aldosterone and cortisol, may have considerable significance for how we view mineralocorticoid receptors in the central nervous system and other non-epithelial tissues where the receptor is unprotected by 11βhydroxysteroid dehydrogenase. Given the very high circulating concentrations of cortisol in the guinea-pig, such unprotected receptors are presumably always occupied by cortisol; the fact that in this species there appears to have been no evolutionary drive toward a lower affinity for cortisol suggests that the physiological action of such receptors is somehow predicated on their being always occupied. Whether, as has been suggested for mineralocorticoid receptors acting at composite response elements,[39] they may have a role in dampening glucocorticoid receptor mediated responses remains to be determined.

A second recent study[59] in the area of mineralocorticoid receptors also underlines the necessity of keeping an open mind in conceptualizing the subcellular actions of mineralocorticoid receptors. When mineralocorticoid receptors and glucocorticoid receptors are cotransfected into neuroblastoma cells with an MMTV-luc reporter, at low cortisol concentrations they appear to have a synergistic effect, with response clearly greater than the sum of the responses when each receptor alone is transfected. When both receptors are incubated with a ^{32}P-labeled MMTV sequence containing a canonical response element, the complex is supershifted by antibodies to either the mineralocorticoid receptor or glucocorticoid receptor. In addition, the dissociation rate of putative mineralocorticoid receptor-glucocorticoid receptor heterodimers from canonical response elements is clearly slower than that of either receptor alone. Though these studies strongly suggest the existence and functional importance of such heterodimers in cotransfection systems, there are currently very few studies that address this question under physiological conditions. There are tissues, both epithelial (colon) and nonepithelial (hippocampus) which express not dissimilar levels of mineralocorticoid and glucocorticoid receptors, and which thus may be suitable to explore the extent to which the phenomena described for multiply transfected neuroblastoma cells are reflected in normal cells *in vivo*.

Third, pathophysiological roles for mineralocorticoid receptors on nonepithelial tissue outside the central nervous system have been suggested by recent studies on the involvement of steroid hormones in the genesis of cardiac fibrosis. In rats sensitized by uninephrectomy and sodium loading, infusion of a modest dose of aldosterone (0.75 µg/h) for 8 weeks produces very marked interstitial collagen deposition in the heart,[60] an effect seen in both ventricles (thus presumably pressure-independent) and able to be blocked by spironolactone given at a dose with minimal blood pressure lowering effects. In subsequent studies[61] perivascular fibrosis, under similar experimental circumstances, was shown to be increased by deoxycorticosterone and RU38486 more than by aldosterone, suggesting a combined effect of mineralocorticoid agonist-glucocorticoid antagonist occupancy of cardiac corticosteroid receptors; whereas aldosterone at high doses has been shown to be a glucocorticoid agonist, deoxycorticosterone is antagonist.[62] The heart has clearly been shown to express mineralocorticoid receptors;[2,63] the extent to which the effects of aldosterone, deoxycorticosterone and the glucocorticoid antagonist RU38486 are mediated directly via effects on cardiac mineralocorticoid and glucocorticoid receptors awaits exploration.

IX. PSEUDOHYPOALDOSTERONISM

Finally, in any consideration of classical mineralocorticoid receptors, the clinical syndrome of pseudohypoaldosteronism deserves some commentary. Patients with pseudohypoaldosteronism commonly present as neonates with hyponatremia, hyperkalemia, and the stigmata of circulatory collapse despite elevated renin and aldosterone levels; from the description of the index case in 1958[64] it was proposed that the syndrome may reflect a deficit in receptor mechanisms for aldosterone. In 1985, studies on the index case and two siblings with the syndrome found absent or very low levels of [^{3}H]aldosterone binding in circulating mononuclear leukocytes compared with normals, apparently confirming a receptor defect as the underlying etiology of the syndrome, as has subsequently been demonstrated for a variety of other steroid and thyroid hormone receptors. Very recently, however, in two separate laboratories the mineralocorticoid receptor sequence has been cloned from the index case,[65] with the sporadic form of the syndrome, and from an unrelated patients,[66] with the familial form. In both instances no sequence abnormality could be detected, suggesting that the abrogation of binding and function reflects abnormalities elsewhere in the admittedly poorly explored area of mineralocorticoid receptor chaperoning and activation.[67] In today's world of molecular genetics, it is perhaps instructive to reflect that in neither apparent mineralocorticoid excess nor pseudohypoaldosteronism, conditions which have proven so instructive in terms of mineralocorticoid receptors and specificity conferring mechanisms, have there been demonstrated any sequence abnormalities in the genes initially presumed to be centrally involved in the clinical conditions.

REFERENCES

1. **Armanini, D., Strasser, T., and Weber, P. C.,** Characterization of aldosterone binding sites in circulating human mononuclear leukocytes, *Am. J. Physiol.,* 248, E388, 1985.
2. **Arriza, J. L., Weinberger, C., Glaser, T., Handelin, B. L., Housman, D. E., and Evans, R. M.,** Cloning of the human mineralocorticoid receptor complementary DNA: structural and functional kinship with the glucocorticoid receptor, *Science,* 237, 268, 1987.
3. **Krozowski, Z. S., and Funder J. W.,** Renal mineralocorticoid receptors and hippocampal corticosterone binding species have identical intrinsic steroid specificity, *Proc. Natl. Acad. Sci. U.S.A.,* 80, 6056, 1983.
4. **Myles, K. and Funder, J. W.,** Type I (mineralocorticoid) receptors in the guinea-pig, *Am. J. Physiol.,* in press 1994.
5. **Gomez-Sanchez, E. P., Venkataraman, M. T., Thwaites, D., and Fort, C.,** Intracerebroventricular infusion of corticosterone antagonizes ICV-aldosterone hypertension, *Am. J. Physiol.,* 258, E649, 1990.
6. **de Kloet, E. R., Sybesma, H., and Reul, J. H. M.,** Selective control by corticosterone of serotonin-1 receptor capacity raphe-hippocampal system, *Neuroendocrinology,* 42, 513, 1986.
7. **Rousseau, G., Baxter, J. D., Funder, J. W., Edelman, I. S., and Tomkins, G. M.,** Glucocorticoid and mineralocorticoid receptors for aldosterone, *J. Steroid Biochem.,* 3, 209, 1972.
8. **Funder, J. W., Feldman, D., and Edelman, I. S.,** Specific aldosterone binding in rat kidney and parotid, *J. Steroid Biochem.,* 3, 209, 1972.
9. **Funder, J. W., Feldman, D., and Edelman, I. S.,** The roles of plasma binding and receptor specificity in the mineralocorticoid action of aldosterone, *Endocrinology,* 92, 994, 1973.
10. **Marver, D., Stewart, J., Funder, J. W., Feldman, D., and Edelman, I. S.,** Renal aldosterone receptors: studies with (3H) aldosterone and the anti-mineralocorticoid (3H) spirolactone (SC-26304), *Proc. Natl. Acad. Sci. U.S.A.,* 71, 1431, 1974.
11. **Feldman, D., Funder, J. W., and Edelman, I. S.,** Evidence for a new class of corticosterone receptors in the rat kidney, *Endocrinology,* 92, 1429, 1973.
12. **McEwen, B. S., Weiss, J. M., and Schwartz, L. S.,** Selective retention of corticosterone by limbic structures in rat brain, *Nature,* 220, 911, 1968.
13. **Nestler, E. J., Rainbow, T. C., McEwen, B. S., and Greengard, P.,** Corticosterone increases the level of protein 1, a neuron-specific protein in rat hippocampus, *Science,* 212, 1162, 1981.
14. **Beaumont, K. and Fanestil, D. D.,** Characterization of rat brain aldosterone receptors reveals high affinity for corticosterone, *Endocrinology,* 113, 2043, 1983.

15. **Wrange, O. and Yu, Z.-Y.,** Mineralocorticoid receptor in rat kidney and hippocampus: characterization and quantitation by isoelectric focusing, *Endocrinology,* 113, 243, 1983.

16. **Sheppard, K. E. and Funder, J. W.,** Cortisol 17β acid, transcortin, and the heterogeneity of rat brain glucocorticoid receptors, *J. Steroid Biochem.,* 25, 285, 1986.

17. **Sheppard, K. and Funder, J. W.,** Mineralocorticoid specificity of renal Type I receptors: in vivo binding studies, *Am. J. Physiol.,* 252, E225, 1987.

18. **Sheppard, K. and Funder, J. W.,** Type I receptors in parotid, colon and pituitary are aldosterone-selective *in vivo, Am. J. Physiol.,* 253, E467, 1987.

19. **Patel, P. D., Sherman, T. G., Goldman, D. J., and Watson, S. J.,** Molecular cloning of a mineralocorticoid (Type I) receptor complementary DNA for rat hippocampus, *Mol. Endocrinol.,* 3, 1877, 1989.

20. **Hollenberg, S. M., Weinberger, C., Ong, E. S., Cerelli, G., Oro, A., Lebo, R., Thompson, E. B., Rosenfeld, M. G., and Evans, R. M.,** Primary structure and expression of a functional human glucocorticoid receptor cDNA, *Nature,* 318, 635, 1985.

21. **Loosfeld, H., Atger, M., Misrahi, M., Guichon-Mantel, A., Meriel, C., Logeat, F., Bnarous, R., and Milgrom, E.,** Cloning and sequence analysis of rabbit progesterone-receptor complementary DNA, *Proc. Natl. Acad. Sci. U.S.A.,* 83, 9045, 1986.

22. **Chang, C., Kokontis, J., and Liao, S.,** Molecular cloning of human and rat complementary DNA encoding androgen receptors, *Science,* 240, 324, 1988.

23. **Ulick, S., Ramirez, L. C., New, M.,** An abnormality in reductive metabolism in a hypertensive syndrome, *J. Clin. Endocrinol. Metab.,* 44, 799, 1977.

24. **Ulick, S., Levine, L. S., Gunczler, P., Zanconato, G., Ramirez, L. C., Rauh, W., Rosler, A., Bradlow, H. L., and New, M. I.,** A syndrome of apparent mineralocorticoid excess associated with defects in the peripheral metabolism of cortisol, *J. Clin. Endocrinol. Metab.,* 49, 757, 1979.

25. **Stewart, P. M., Shackleton, C. H. L., and Edwards, C. R. W.,** The cortisol-cortisone shuttle and the genesis of hypertension, in *Corticosteroids and Peptide Hormones in Hypertension,* Vol. 39, Serono Symposia Publications, Mantero, F. and Vecsei P., Eds., Raven Press, New York, 1981, 163.

26. **Stewart, P. M., Wallace, A. M., Valentino, R., Burt, D., Shackleton, C. H. L., and Edwards, C. R. W.,** Mineralocorticoid activity of liquorice: 11-beta-hydroxysteroid dehydrogenase deficiency comes of age, *Lancet,* 2, 821, 1987.

27. **Funder, J. W., Pearce, P. T., Smith, R., and Smith, A. I.,** Mineralocorticoid action: target-tissue specificity is enzyme, not receptor-mediated, *Science,* 242, 583, 1988.

28. **Edwards, C. R. W., Stewart, P. M., Burt, D., McIntyre, M. A., de Kloet, E. R., Brett, L., Sutano, W. S., and Monder, C.,** Localization of 11β-hydroxysteroid dehydrogenase — tissue specific protector of the mineralocorticoid receptor, *Lancet,* 2, 986, 1988.

29. **Lakshmi, V. and Monder, C.,** Purification and characterization of the corticosteroid 11β-dehydrogenase component of the rat liver 11β-hydroxysteroid dehydrogenase complex, *Endocrinology,* 123, 2390, 1989.

30. **Agarwal, A. K., Monder, C., Ecksten, B., and White, P. C.,** Cloning and expression of a rat cDNA encoding corticosterone 11β dehydrogenase, *J. Biol. Chem.,* 264, 18939, 1989.

31. **Rundle, S. E., Funder, J. W., Lakshmi, V., and Monder, C.,** The intrarenal localization of mineralocorticoid receptors and 11β dehydrogenase: immunocytochemical studies, *Endocrinology,* 125, 1700, 1989.

32. **Funder, J. W.,** 11β hydroxysteroid dehydrogenase and the meaning of life, *Mol. Cell. Endocrinol.,* 68, C3, 1990.

33. **Oberfield, S. E., Levine, L. S., Carey, R. M., Greig, F., Ulick, S., and New, M. I.,** Metabolic and blood pressure responses to hydrocortisone in the syndrome of Apparent Mineralocorticoid Excess, *J. Clin. Exp. Endocrinol.,* 56, 332, 1983.

34. **Armanini, D., Kuhnle, U., Strasser, T., Dorr, H., Weber, P. C., Stockigt, J. R., Pearce, P., and Funder, J. W.,** Aldosterone-receptor deficiency in pseudohypoaldosteronism, *N. Engl. J. Med.,* 313, 1178, 1985.

35. **Funder, J. W., Pearce, P. T., Myles, K., and Roy, L. P.,** Apparent mineralocorticoid excess, pseudohypoaldosteronism and urinary electrolyte excretion: towards a redefinition of "Mineralocorticoid" action, *FASEB J.*, 4, 3234, 1990.
36. **Naray-Fejes-Toth, A. and Fejes-Toth, G.,** Glucocorticoid receptors mediate mineralocorticoid-like effects in cultured collecting duct cells, *Am. J. Physiol.*, 259, F672, 1990.
37. **Stewart, P. M.,** The effect of 11β-OHSD on mineralocorticoid and glucocorticoid gene transcription, *Steroids*, 59, 90, 1994.
38. **Pearce, D. and Yamamoto, K. R.,** Mineralocorticoid and glucocorticoid receptor activities distinguished by nonreceptor factors at a composite response element, *Science*, 259, 1661, 1993.
39. **Funder, J. W.,** Mineralocorticoids, glucocorticoids, receptors and response elements, *Science*, 259, 1132, 1993.
40. **Adler, A. J., Danielsen, M., and Robins, D. M.,** Androgen-specific gene activation via consensus glucocorticoid response element is determined by interaction with nonreceptor factors, *Proc. Natl. Acad. Sci. U.S.A.*, 89, 11660, 1992.
41. **Castren, M. and Damm, K.,** A functional promoter directing expression of a novel type of rat mineralocorticoid receptors mRNA in brain, *J. Neuroendocrinol.*, 5, 461, 1993.
42. **Doyle, D., Krozowski, Z. S., Morgan, F. J., and Funder, J. W.,** Analysis of renal and hippocampal type I and type II receptors by fast protein liquid chromatography, *J. Steroid Biochem.*, 29, 611, 1988.
43. **Nikkila, H., Tannin, G. M., Taylor, N. F., Kalaitzoglou, G., Monder, C., and White, P. C.,** Defects in the HSD11 gene encoding 11β-hydroxysteroid dehydrogenase are not found in patients with apparent mineralocorticoid excess or 11-oxoreductase deficiency, in *Proc. 7th Ann. Meet. Endocr. Soc.*, Las Vegas, Abstr. 410B, 153, 1993.
44. **Mantero, F.,** Apparent mineralocorticoid excess Type II, *Steroids*, 59(2), 80, 1994.
45. **Hierholzer, K.,** Endogenous inhibitors of 11β-OHSD, *Steroids*, 59(2), 131, 1994.
46. **Monder, C.,** Comparative aspects of 11β-OHSD, *Steroids*, 59(2), 69, 1994.
47. **Duperrex, H., Kenouch, S., Gaeggeler, H. P., Seckl, J. R., Edwards, C. R., Farman, N., and Rossier, B. C.,** Rat liver 11 beta-hydroxysteroid dehydrogenase complementary deoxyribonucleic acid encodes oxoreductase activity in a mineralocorticoid-responsive toad bladder cell line, *Endocrinology*, 132, 612, 1993.
48. **Naray-Fejes-Toth, A.,** 11β-OHSD in aldosterone target cells: new enzymes, old protein, *Steroids*, 59, 105, 1994.
49. **Mercer, W. R. and Krozowski, Z. S.,** Localization of an 11β-hydroxysteroid dehydrogenase activity to the distal nephron of the rat kidney. Evidence for the existence of two species of dehydrogenase in the rat kidney, *Endocrinology*, 130, 540, 1992.
50. **Brown, R. W., Chapman, K. E., Edwards, C. R. W., and Seckl, J. R.,** Human placental 11β-hydroxysteroid dehydrogenase: evidence for and partial purification of a distinct NAD-dependent isoform, *Endocrinology*, 132, 2614, 1993.
51. **Yang, K., Smith, C. L., Dales, D., Hammond, G. L., and Challis, J. R. G.,** Cloning of an ovine 11β-hydroxysteroid dehydrogenase complementary deoxyribonucleic acid: tissue and temporal distribution of its messenger ribonucleic acid during fetal and neonatal development, *Endocrinology*, 131, 2120, 1992.
52. **Rupprecht, R., Reul, J. M. H. M., van Steensel, B., Spengler, D., Soder, M., Berning, B., Holsboer, F., and Damm, K.,** Pharmacological and functional characterization of human mineralocorticoid and glucocorticoid receptor ligands, *Eur. J. Pharmacol.*, 247, 145, 1993.
53. **Hierholzer, K., Schoneshofer, M., Siebe, H., Tsiakiras, D., and Weskamp, P.,** Corticosteroid metabolism in isolated rat kidney in vitro. I. Formation of lipid insoluble metabolites from corticosterone (B) in renal tissue from male rats, *Pflugers Arch.*, 400, 363, 1984.
54. **Funder, J. W.,** A la recherche du temps perdu, in *Aldosterone: Fundamental Aspects*, Bonvalet, J.-P., Farman, N., Lombes, M., and Rafestin-Oblin, M.-E., Eds., John Libbey Eurotext Limited, France, 1991, 77.
55. **Smith, A. I., Wallace, C. A., Moritz, R. L., Simpson, R. J., Schmauk-White, L. B., Woodcock, E. A., and Funder, J. W.,** Isolation, amino acid sequence and superagonist steroidogenic activity of guinea-pig ACTH, *J. Endocrinol.*, 115, R5, 1987.

56. **Westphal, U.,** Steroid-protein interactions. XIII. Concentration and binding affinities of corticosteroid binding globulins in sera of man, monkey, rat, rabbit and guinea-pig, *Arch. Biochem. Biophys.,* 118, 556, 1967.
57. **Kraft, N., Hodgson, A. J., and Funder, J. W.,** Glucocorticoid receptor and effector mechanisms: a comparison of the corticosensitive mouse with the corticoresistant guinea-pig, *Endocrinology,* 104, 344, 1979.
58. **Keightley, M.-C. and Fuller, P. J.,** Unique sequences in the guinea-pig glucocorticoid receptor induce constitutive transactivation and decrease steroid sensitivity, *Mol. Endocrinol.,* in press, 1994.
59. **Trapp, T., Rupprecht, R., and Holsboer, F.,** Coexpression of the mineralocorticoid receptor and glucocorticoid receptor: characterization of transactivation and DNA-binding properties, *J. Cell. Biochem.,* Suppl. 18B, Abstr. K139, 349, 1994.
60. **Brilla, C. G. and Weber, K. T.,** Mineralocorticoid excess, dietary sodium and myocardial fibrosis. *J. Lab. Clin. Med.,* 120, 893, 1992.
61. **Young, M., Fullerton, M., Dilley, R., and Funder, J.,** Mineralocorticoids, hypertension and cardiac fibrosis, *J. Clin. Invest.,* in press, 1994.
62. **Marks, R., Barlow, J., and Funder, J. W.,** The mechanisms of steroid-induced vasoconstriction: glucocorticoid antagonist studies, *J. Clin. Endocrinol. Metab.,* 54, 1075, 1982.
63. **Pearce, P. and Funder, J. W.,** High affinity aldosterone binding sites (Type I receptors) in rat heart, *Clin. Exp. Pharmacol. Physiol.,* 14, 859, 1987.
64. **Cheek, D. B. and Perry, J. W.,** A salt wasting syndrome in infancy, *Arch. Dis. Child.,* 33, 252, 1958.
65. **Komesaroff, P. A., Verity, K., and Fuller, P. J.,** Pseudohypoaldosteronism: molecular characteristion of the mineralocorticoid receptor, *J. Clin. Endocrinol. Metab.,* in press, 1994.
66. **Zennaro, M.-C., Borensztein, P., Jeunemaitre, X., Armanini, D., and Soubrier, F.,** No alteration in the primary structure of the mineralocorticoid receptor in a family with Pseudohypoaldosteronism, *J. Clin. Endocrinol. Metab.,* in press, 1994.
67. **Corvol, P. and Funder, J. W.,** The enigma of pseudohypoaldosteronism, *J. Clin. Endocrinol. Metab.,* in press, 1994.
68. **Myles, K. and Funder, J. W.,** unpublished observations.

Chapter 2

GENOMIC EFFECTS OF ALDOSTERONE ON THE KIDNEY: CELLULAR AND MOLECULAR ASPECTS

Catherine Barlet-Bas and Alain Doucet

Table of Contents

17

I. INTRODUCTION

In Metazoa with individualized and specialized organs, information needs to circulate between the different organs to allow their coordinated functioning necessary for survival of the whole organism. One of the communication pathways between cells is ensured by hormones which are chemical messages synthesized by specialized cells in response to stimuli and released and carried in the blood flow to their target cells. The specificity of target cells is accounted for by the presence of receptors which are proteins that specifically bind the hormones and transduce their message within cells. Steroid hormones are derivatives of cholesterol which have been well conserved during evolution. Aldosterone is the main mineralocorticosteroid which, in mammals and in particular in humans, plays an important role in the regulation of sodium homeostasis and blood pressure.

Aldosterone was isolated from extracts of adrenal cortex more than 20 years after other corticosteroids. Concomitantly, clinicians isolated from the urine of patients with abnormal sodium retention a substance which later proved to be aldosterone,[1] a first indication that aldosterone might be involved in the regulation of sodium handling. Later, it was shown that injection of aldosterone in dog renal artery altered the electrolytic composition of their urine, thereby identifying the kidney as a target for aldosterone.[2] Aldosterone decreases urinary sodium excretion (antinatriuretic effect) and increases that of potassium (kaliuretic effect) and of protons.[2-5] Stop-flow experiments revealed quite early that the distal parts of the nephron were the sites of action of aldosterone.[6]

Some 30 years ago, Edelman and colleagues[7] proposed a model for the action of aldosterone on renal sodium transport which has been verified and further elaborated by many experiments since then. Aldosterone freely diffuses through epithelial cell membranes and binds to cytoplasmic receptors which are thereby activated and translocated to the cell nucleus where they bind to the DNA. In turn, this controls the synthesis of specific mRNAs encoding for proteins which account for the cellular effects of the hormone. It is now estimated that aldosterone controls the synthesis (up- or down-regulation) of several hundreds distinct proteins in target cells.[8,9] These aldosterone-induced (or -repressed) proteins may be involved either in constitutive pathways for cation transport (Na^+, K^+, H^+) or in regulatory pathways.

In this chapter, we shall successively consider: (1) the general features of the genomic actions of aldosterone, including considerations on the mineralocorticoid receptor and on the interaction between aldosterone, its receptor and DNA; (2) the biologic effects of aldosterone on the renal hydroelectrolytical metabolism, and (3) the molecular characterization of aldosterone-induced proteins in renal cells.

II. INTERACTIONS BETWEEN ALDOSTERONE, ITS RECEPTOR AND DNA

A. GENERAL STRUCTURE OF STEROID RECEPTORS

In contrast to other receptors for steroid hormones (estrogen receptor, ER; progesterone receptor, PR; and glucocorticoid receptor, GR), which have been well characterized and purified many years ago, the properties of the mineralocorticoid receptor (MR) are not fully characterized as yet. This delay may be accounted for by the low concentration of receptors in target cells, by their lability, and by the complexity of the system since, in mammals, both mineralo- and glucocorticoids bind to the MR.

Cytoplasmic aldosterone binding sites were demonstrated for the first time by Herman et al. using *in vitro* labeling with tritiated aldosterone in rat kidney.[10] Based on their sensitivity to proteolytic enzymes, it was proposed that these binding sites were proteins. Furthermore, these authors reported that estradiol had no affinity for these receptors whereas 9α-fluorocortisol and deoxycorticosterone inhibited the binding of tritiated aldosterone. More recent studies demonstrated that aldosterone binds to several types of receptors (see chapter 1), but we shall focus the following discussion on the high affinity receptors for aldosterone, i.e., the Type I receptors (affinity: 0.1 to 1 n*M*, number of sites: 3 to 200 fmol/mg protein, according to the tissue and species).

Many studies based on either immunological methods, molecular biology, or heterologous expression systems made it possible to demonstrate that steroid receptors consist of three major distinct domains; an immunogenic domain, a DNA binding domain, and a hormone binding domain (Figure 1). This basic structure applies to all the members of the steroid receptor superfamily, suggesting that they are derived from a common ancestor.

The DNA binding domain, encoded by the C domain, is the best conserved among different steroid receptors (for example, 94% homology between MR and GR). It consists of about 70 amino acids and contains "zinc finger" structures in which cysteines are coordinated by atoms of zinc. The first finger, or CI (located on the NH_2 side), which contains four cysteines and several hydrophobic amino acids, is responsible for the specificity of the hormone response. The second finger, or CII (located on the COOH side) contains five cysteine residues and several basic amino acids. This region is likely involved in the protein-protein interactions responsible for the dimerisation of the receptor.[11] The tridimensional structure of the DNA binding domain of the GR reported by Härd et al.[12] demonstrated the presence of two successive zinc finger structures, each one consisting of an α helix followed by a closeby linear region. These α helices are perpendicular and constitute, with the closeby linear regions, a globular structure from which the zinc fingers emerge. The DNA binding domain recognizes specific target sequences on the DNA, called hormone response elements (HREs). The GR, PR, androgen receptor (AR) and

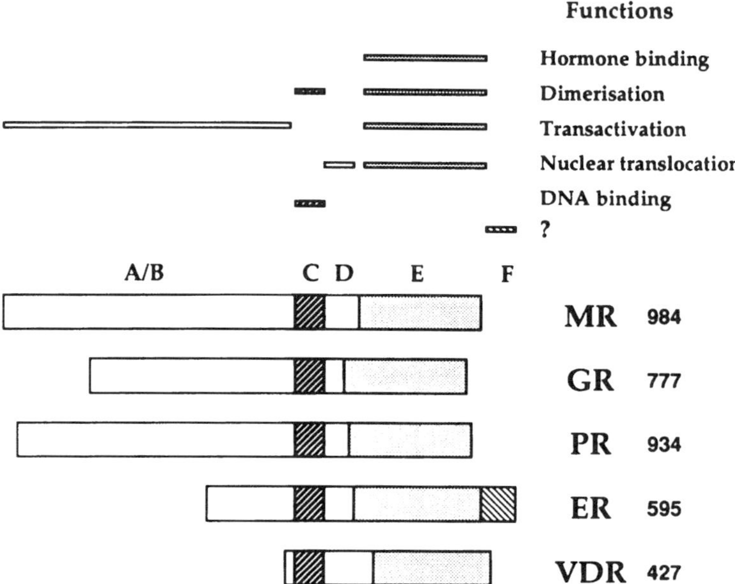

FIGURE 1:   Schematic representation of the organization of steroid receptors (MR, mineralo-corticoid receptor; GR, glucocorticoid receptor; PR, progesterone receptor; ER, estrogen receptor; VDR, vitamin D receptor). The receptors consist of six distinct regions (A-F) with specific functions. The number of amino acids (indicated next to the abbreviation for each receptor) constituting the receptor molecule mainly varies according to the length of the A/B and F domains. The function of the F domain, which is exclusively found in the estrogen receptor, remains unknown. Redrawn with permission from Evans.[11]

MR, which have very similar DNA binding domains, recognize the same response elements, the glucocorticoid response (GREs), the consensus sequence of which is 5′-NGGTACANNNTGTTCTN-3′ (where N represents any nucleotide). This con-sensus sequence is palindromic so that receptors can bind symmetrically to DNA, which suggests that the receptor binds to DNA under a dimeric configuration.[11]

The hormone binding domain is linked to the DNA binding domain by a hinge region encoded by the D domain, the role of which is not yet fully understood (Figure 1). The hormone binding domain is encoded by the E domain, a well conserved region among distinct steroid receptors (57% homol-ogy between MR and GR). It consists of 250 amino acids forming a hydropho-bic pocket which needs to be intact to allow not only the hormone binding but also the dimerisation of the receptor, its nuclear translocation, and the resulting activation of the transcription.[13-16] This region also contains the binding site for the heat shock protein (hsp 90), a chaperone protein associated with the unbound receptor which facilitates the hormone response.[17]

The N-terminus domain is highly variable between steroids, and for a given receptor, between species. For example, in humans, there is less than 15% homology between these domains of the MR on the one hand and of the GR on the other. The size of this domain varies from 25 to 602 amino acids for the vitamin D receptor and for the MR, respectively. This hypervariable region contains the epitopes of most anti-receptor antibodies and is therefore called the immunogenic domain. Along with the hormone binding domain, it contains a region controlling transcription. Functional characterization of artificial chimeras constructed with different domains of distinct receptors demonstrated that each domain can work independently of the others.

B. SPECIFIC FEATURES
OF THE MINERALOCORTICOID RECEPTOR

The cDNA encoding for the human MR was cloned in 1987 by Arriza,[18] who took advantage of the homology between the DNA binding regions of MR and GR to isolate a portion of human genomic DNA which he later used as a probe to obtain the full length cDNA. The expression product of this cDNA binds aldosterone with high affinity and functions as a transcription factor, as demonstrated on the promoter of the Mouse Mammary Tumor Virus (MMTV).

The gene of the human MR encodes for a 984 amino acid protein (107 kDa) which includes several domains as is the case for other steroid receptors. It is associated to hsp 90, probably under a dimeric form. During the transformation step, i.e., the passage from a form of the MR that does not bind to DNA to one that binds to it, the sedimentation coefficient of the receptor decreases from 9S to 4S at the same time as hsp 90 dissociates.[19] However, Alnemri et al.[20] have reported that, when over-expressed in the baculovirus system, the non-transformed MR (9S) could also bind to DNA cellulose and this interaction was not modified by *in vitro* activation (by heating or by increasing the ionic strength). The authors concluded that the dissociation between hsp 90 and the MR was not an obligatory step for the binding to DNA, suggesting that hsp 90 might play a different role with GR and MR. Conversely, studies by Schulman et al.[21] on rat colon failed to demonstrate the binding of the free MR to DNA cellulose after its heat activation. Altogether, these results reveal differences in the activation process of the MR from one experimental model to another, and the precise role of hsp 90 in the mechanism of action of the free MR remains to be fully elucidated.

C. INTERACTION BETWEEN MR AND DNA

Through binding to hormone responsive elements (HREs) on DNA, the aldosterone-MR complex acts as a transcriptional factor that controls the expression of specific genes. Due to the high degree of homology between their DNA binding domains, MR and GR bind with high affinity to the same consensus sequences of the DNA, the GREs, close to hormone responsive promoters.[13] For example, human MR and GR control the same promoters such as those of MMTV, of tyrosine amino transferase, and of tryptophane oxygenase.[22] This raises the question of how glucocorticoids and mineralocorticoids may induce distinct responses in a same cell. To explain steroidal specificity, Pearce and Yamamoto[23] have proposed four hypotheses: (1) the presence of a single class of receptors in a given cell type; (2) the possibility that target cells inactivate one of the two types of hormones; this hypothesis probably accounts for the specificity of aldosterone action on sodium transport in the collecting duct since its cells express both MR and GR, but also a 11βhydroxysteroid dehydrogenase (11βHSD) activity that degrades glucocorticoids (see Chapter 1); (3) the presence, along with the GREs, of other DNA sequences that can interact specifically with either MR or GR; and (4) the possibility for hormone-receptor complexes to interact with other transcriptional factors. Indeed, composite response elements have been identified, which are 25 nucleotides long sequences displaying binding sites for both the receptor and nonreceptor factors essential for the activity of the receptor.[24,25] For example, Pearce and Yamamoto recently reported that, although MR and GR had the same efficiency on GRE, they induced distinct actions on plfG, a low affinity HRE that binds the AP1 transcription factor. Under conditions where the GR repressed the AP1-induced transcription of plfG, the MR was inactive. Using MR-GR chimeras, these authors demonstrated that part of the N-terminus of the GR (amino acids 105 to 438) was necessary for the repressive effect of GR on plfG, suggesting that the specificity of glucocorticoid action might be accounted for in part by the interaction between this N-terminus portion of the GR and accessory, nonreceptor factors. Finally, it should be noted that the steroid specificity might also be provided by interaction with intracellular mechanisms triggered by aldosterone binding to membrane receptors (see Chapter 5).

In summary, the regulation pattern of a given response element might be determined by the ligand availability, the expression of 11βHSD and the presence of nonreceptor factors (such as AP1) interacting with the receptor.[26]

III. EFFECTS OF ALDOSTERONE ON KIDNEY FUNCTION

In vertebrates, and particularly in mammals, the kidneys play an essential role in the homeostasis of extracellular fluids (volume and composition), and accordingly they are targets for many regulatory factors including hormones. Aldosterone is one of the main hormones controlling sodium, potassium, and hydrogen balance of the organism.[27,28] Before analyzing the effects of aldosterone on ion transport, it is important to recall the main characteristics of aldosterone target cells in kidney and the experimental models allowing their study.

A. EXPERIMENTAL MODELS

The effects of aldosterone on renal handling of solutes and water have been first assessed by clearance studies in which the changes in urine volume and composition induced by administration of aldosterone to normal or adrenalectomized animals were recorded. However, the mammalian kidney is an heterogeneous organ constituted by numerous units — the nephrons — contributing in parallel to the formation of urine from the glomerular ultrafiltrate. The nephrons — which are tubular monolayers of epithelial cells a few centimeters of length and 5 to 50 μm of diameter — are themselves heterogeneous structures, made of successive segments displaying specific structural, biochemical, and functional properties (Figure 2). In the kidney, aldosterone acts mainly (if not exclusively), on the collecting duct which is a tight epithelium, i. e., an epithelium displaying a high electrical resistance. Therefore, the study of the renal action of aldosterone has required the development of specific methods allowing one to circumvent the axial heterogeneity of the nephron, and to study the transport properties and the regulation of these specific segments. These include, on the one hand, *in vivo* micropuncture and microperfusion experiments and, on the other, microdissection of well-defined nephron segments[31] in combination with *in vitro* tubular microperfusion[32] or biochemical microassays (reviewed in Reference 33).

Two types of *in vitro* experiments were performed to study the action of aldosterone on tubular transport. In most experiments, the plasma level of aldosterone was manipulated *in vivo* (by adrenalectomy, acute or chronic administration of aldosterone, or change of the cation content of the diet), and the transport function of the tubules was determined *in vitro* after their isolation by microdissection. This approach supposes that the cells keep the "memory" of their *in vivo* steroidal environment, which is likely for hormones acting through induction of protein synthesis. In some studies, investigators have evaluated the effects of aldosterone added *in vitro* on nephron segments dissected from either normal or mineralocorticoid-deficient animals.

It should be stressed, however, that the collecting duct, i.e., the main renal target of aldosterone, is heterogeneous since it consists of at least three distinct intermingled cell types (Figure 2): principal cells, responsible for sodium and water reabsorption and potassium secretion; type A intercalated cells, responsible for proton secretion and bicarbonate and potassium reabsorption; and type B intercalated cells, responsible for proton and chloride reabsorption and bicarbonate secretion (reviewed in Reference 30). A schematic representation of membrane ion transport systems responsible for these ion transports in the different cell types constituting the collecting duct is given in Figure 2.

Experimental models of tight epithelia, such as the frog skin or the urinary bladder of amphibians or of the turtle, have also been very useful to study the action of aldosterone, even though data derived from these models are not always directly transposable to the mammalian nephron. These planar tight epithelia can be mounted in Ussing's chambers for flux and electrophysiological measurements. According to the model of Ussing et al.,[34,35] the flux of sodium across these epithelia can be easily determined by the short-circuit current measured when the apical and basolateral faces of the epithelia are bathed with symmetrical solutions. The main limitation of these experimental models for the study of aldosterone action results from their much lower

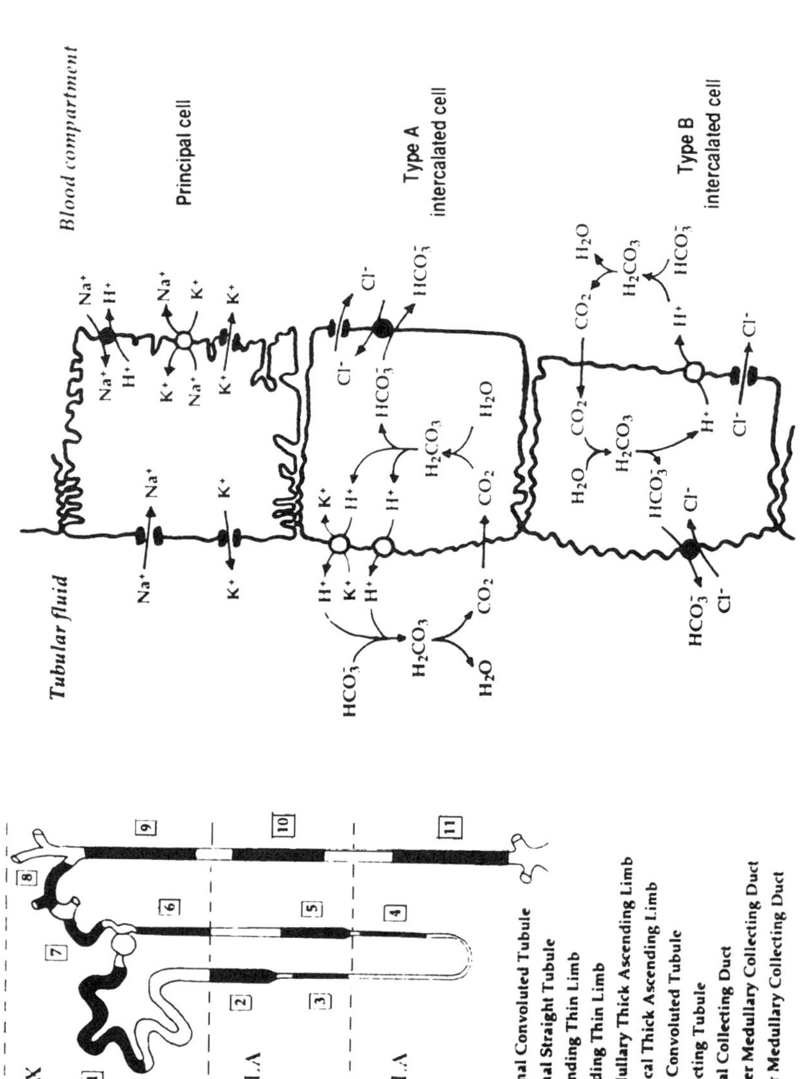

Blood compartment
Tubular fluid
Principal cell
Type A intercalated cell
Type B intercalated cell
Na+
H+
Na+
K+
K+
Na+
H+
K+
Na+
K+
Na+
Na+
K+
Cl-
HCO3-
Cl-
HCO3-
H2CO3
H2O
CO2
K+
H+
H+
H+
K+
H+
H2CO3
HCO3-
H2O
CO2
CO2
H2O
H2CO3
HCO3-
H2O
H2CO3
H+
H+
HCO3-
Cl-
Cl-
Cl-
HCO3-
Cl-
CORTEX
OUTER MEDULLA
INNER MEDULLA
1
2
3
4
5
6
7
8
9
10
11
PCT : Proximal Convoluted Tubule
PST : Proximal Straight Tubule
DTL : Descending Thin Limb
ATL : Ascending Thin Limb
MTAL : Medullary Thick Ascending Limb
CTAL : Cortical Thick Ascending Limb
DCT : Distal Convoluted Tubule
CNT : Connecting Tubule
CCD : Cortical Collecting Duct
OMCD : Outer Medullary Collecting Duct
IMCD : Inner Medullary Collecting Duct
1 2 3 4 5 6 7 8 9 10 11

FIGURE 2: (Facing page) Organization of the mammalian nephron. **Left:** The dark portions indicate the main segments of nephron (named according to the standard nomenclature recently adopted[29]). **Right:** Schematic (and simplified) representation of membrane ion transporters in the different types of cells constituting the cortical (CCD) and outer medullary collecting duct (OMCD) (reviewed in Reference 30). In both CCD and OMCD, principal cells account for 60 to 65% of the cell population in both rat and rabbits, whereas intercalated cells account for the remaining 35 to 40%. Among intercalated cells, the proportion of type A cells increases from about one third at the beginning of the CCD to 100% at the end of the OMCD. *Principal cells:* Sodium reabsorption proceeds in a two-step mechanism which includes pumping of Na^+ out of the cells by the basolateral Na,K-ATPase which generates a driving force for apical Na^+ entry via amiloride-sensitive sodium channels. Na,K-ATPase also accumulates K^+ into the cells, where it leaks through either apical or basolateral potassium channels. The apical membrane depolarization induced by conductive sodium entry favors secretion of K^+ into the duct lumen over recycling across the basolateral membrane. The basolateral membrane is also equipped with a Na/H antiporter involved in the regulation of intracellular pH. It should be stressed that the sodium flux through this carrier is negligible as compared to the sodium flux via the apical channels. *Type A intercalated cells:* Protons are primarily pumped out of this cell type across the apical membrane by the electrogenic H-ATPase and, to a smaller extent, by an electroneutral H,K-ATPase. Protons are either excreted in the urine as such (which lowers urinary pH) or combined with bicarbonate, giving rise to CO_2 which diffuses into the cell through the apical membrane. Within the cell, where pH is more alkaline (due to proton extrusion), CO_2 regenerates protons and bicarbonate, the latter of which is exchanged against Cl^- across the basolateral membrane. Chloride accumulated within the cell by this process is recycled across the basolateral membrane by chloride channels. *Type B intercalated cells:* These cells are equipped with a basolateral proton pump and an apical bicarbonate/chloride antiporter which allows net bicarbonate secretion. The chloride accumulated into the cell at the apical border leaks through basolateral channels, so that there is a net reabsorption of chloride. The net flux of bicarbonate (reabsorption of secretion) across the collecting duct results from the relative numbers and activities of types A and B intercalated cells.

capacity to reabsorb sodium, as compared to the renal collecting duct. This has been circumvented, in part, by developing cell lines such as A6 cells derived from amphibian kidney[36,37] which can be mounted in Ussing's chambers when grown on filters.

B. EFFECTS ON SODIUM TRANSPORT

Although it had been recognized for many years that patients with Addison's disease display an impaired renal capacity to reabsorb sodium, only in 1961 was a direct stimulatory effect of aldosterone on sodium transport demonstrated by Crabbé.[4] In his seminal report, he showed that aldosterone increased the short-circuit current across the bladder of toad. Later studies have further characterized this stimulation:[38] It appears after a 45 to 90 min latency, followed by an early response lasting for about 3 h, during which sodium transport nearly doubles, whereas the transepithelial electrical resistance decreases by about half. During the following late response, which extends over 12 to 24 h, sodium transport keeps on increasing whereas the electrical resistance is not modified further.

The early response is due to an increase in the rate of apical sodium entry into the cells, since it is mimicked by addition of the sodium ionophore amphotericin B at the apical pole of the cells,[39] an effect which is not additive with that of aldosterone.[40] More recently, aldosterone was reported to increase 3-fold the conductance for sodium of the apical membrane while sodium transport rate increased 2.7-fold.[41] The late response would be accounted for by appearance of new Na,K-ATPase units in the basolateral membrane.[42]

In mammals, the effect of aldosterone on the reabsorption of sodium by the renal tubule has been more difficult to characterize, probably because it depends critically on the experimental conditions. First, the concentration of aldosterone must be within the physiological range (10^{-10} M - 10^{-8} M), because higher concentrations may induce glucocorticoid effects through binding to heterologous receptors. Second, the acute effects of aldosterone should be distinguished from the chronic ones because the latter may result from secondary associated changes of other parameters such as extracellular volume expansion, alteration of tubular sodium delivery, or even morphological and functional adaptations of tubular cells.

The renal action of aldosterone was reevaluated in the light of these considerations in two reports by Horisberger et al. on anesthetized adrenalectomized, glucocorticoid-supplemented rats[43] and by Campen et al.[44] on conscious adrenalectomized rats. They used low doses of aldosterone and studied the time-course of the response: the reduction of urinary sodium excretion occurred 30 to 60 min after aldosterone injection[43] and persisted for several hours. Antinatriuresis reached its maximum with less than 0.3 µg/100 g body weight (b.w.) of aldosterone, which corresponds to a plasma concentration of hormone of approximately 5 nM. Aldosterone induced antinatriuresis in the presence as well as in the absence of glucocorticoids. Curiously, the magnitude of the effect was directly related to the rate of sodium excretion before aldosterone administration. Antinatriuresis was inhibited by the MR antagonist spirono-lactone[44] and by inhibitors of transcription such as actinomycin D.[45-47] This latter effect confirms that aldosterone induces the synthesis of proteins which account for its antinatriuretic effect (see Section IV).

Most studies at the level of *in vitro* microperfused CCD evaluated the effect of long-term corticosteroid administration on transport functions. The first evidence of a long-term action of mineralocorticoids on CCD was reported by Gross et al.[48] who showed that in CCD microdissected from rabbit treated with deoxycorticosterone acetate (DOCA) and perfused *in vitro,* the transepithelial potential difference (PD) was higher than in those from mineralocorticoid-deficient animals. Later it was reported that an increment of sodium transport accompanied this enhanced transepithelial PD.[49-51] Unexpectedly, these changes in transepithelial PD and sodium reabsorption appeared only after several days of treatment with DOCA, peaking after 4 to 18 d.[49,52] When the plasma concentration of aldosterone was varied within a more physiological range by modifying the cation content of the diet, both the transepithelial PD and the sodium reabsorption flux measured *in vitro* were correlated with the *in vivo* aldosterone concentration: Half-maximal increases in voltage and sodium transport were observed at 0.2 to 0.5 nM plasma aldosterone, which corresponds to the K_m value MR for aldosterone. However, here again, these changes reflect chronic action of aldosterone which are correlated with mor-phological changes of the principal cells of the CCD[53] and do not correspond to the short-term (30 min to 1 h) effect of the hormone.

In contrast with the general agreement around the long-term effects of aldosterone on kidney, experiments aimed at eliciting short-term effects by adding aldosterone to *in vitro* microperfused CCD are few and controversial. Gross and Kokko first reported that after a 10 to 20 min latency following the addition of aldosterone to CCD from adrenalectomized rabbits, the transepithelial PD increased from near-zero control values to values in the -20 mV range after 1.5 to 2 h.[54] Using a similar protocol for CCD from rabbits with normal plasma aldosterone level, Schwartz and Burg[50] failed to observe any effect of aldoste-rone added *in vitro* on either transepithelial PD or sodium transport. This apparent discrepancy suggests that the latency before observing a stimulation of sodium transport by aldosterone depends on the aldosterone status of the animals from which CCD are obtained (adrenalectomized vs. normal rabbits) and/or on the basal transport capacity of the CCD before *in vitro* addition of hormone. Wingo et al. reported more recently that *in vitro* addition of aldo-sterone to CCD from adrenalectomized rabbits enhanced sodium reabsorption without altering the transepithelial PD.[55]

Effects of aldosterone on the tubular transport of sodium by nephron seg-ments other than the CCD remain controversial. In the rabbit, aldosterone appears ineffective in both the distal convoluted tubule[48,54] and the medullary collecting duct.[51] Conversely, in the rat nephron, aldosterone has been reported to increase the transepithelial PD and/or the sodium reabsorption in the early and late distal convoluted tubule[56] as well as in the inner medullary collecting duct.[57] Whether these discrepancies reflect differences due to species or to experimental protocol is not clearly established at present.

C. EFFECTS ON POTASSIUM TRANSPORT

It is also well established that Addison's disease is most often associated with hyperkalemia, and that patients with hyperaldosteronism tend to display a negative potassium balance. However, whether the kaliuretic effect of aldosterone is a direct action of the hormone on tubular potassium transport or is secondary to alterations of other associated parameters is still a subject of debate. Indeed, early experiments aimed at evaluating the effect of mineralocorticoid administration on urinary potassium excretion led to controversial results (reviewed in Reference 58).

The issue which will be discussed in order to consider the possible molecular mechanisms accounting for aldosterone effect on potassium transport, focuses on whether the kaliuretic action of aldosterone is entirely secondary to its antinatriuretic action or whether there may be some uncoupling between regulation of sodium and of potassium transport. Here again, the results may differ between the short-term vs. the long-term effects of aldosterone.

Most studies at the level of isolated CCD have evaluated the long-term effect of mineralocorticoid on potassium transport and reported a stimulation.[49,59] For example, in the already mentioned study in which plasma aldosterone level was varied by manipulating the cation content of the diet, Schwartz and Burg reported that not only the rate of sodium reabsorption but also that of potassium secretion measured *in vitro* in isolated rabbit CCD were correlated in a dose-dependent fashion with the *in vivo* aldosterone plasma levels.[50] Stokes carefully examined by means of *in vitro* microperfusion the relationship between sodium reabsorption and potassium secretion in CCD from normal rabbits and from rabbits given DOCA for 1 to 7 d.[59] Under these conditions, sodium reabsorption varied in a wide range (see above) and a linear relationship was observed between the rates of sodium reabsorption and those of potassium secretion in the same tubules. The calculated sodium over potassium flux ratio was 1.34, which is close to the 1.5 stoichiometry of Na,K-ATPase. These data suggest that in the long-term mineralocorticoids alter sodium and potassium transport in parallel, probably through a single mechanism.

In the short-term, the results are more controversial, as they suggest that under some circumstances the effects of aldosterone on sodium and potassium transport may be dissociated: (1) some reports indicated that aldosterone increased potassium transport[32,43] whereas other failed to demonstrate this effect[58,60,61] even though increased sodium reabsorption was observed in all studies; (2) although it is well established that actinomycin D reduces aldosterone-induced changes in sodium transport,[47] some reports indicated that it had the same action on aldosterone-induced kaliuresis[47] whereas others found no effect.[45,62] In fact, several factors (including the dietary potassium intake, the glomerular filtration rate, the acid/base balance, and the transepithelial PD and/or the sodium reabsorption rate in the distal nephron segments) are known to influence the overall potassium excretion and may therefore modulate the action of aldosterone on the collecting duct. For example, stimulation of potassium excretion by aldosterone is better evidenced in potassium-depleted animals than in normal ones[46,63] and, adequate sodium supply to the distal nephron is required for observing mineralocorticoid-induced kaliuresis.[64,65]

In summary, it seems likely that the long-term effects of aldosterone on sodium reabsorption and potassium secretion along the collecting duct are tightly coupled. In the short-term, aldosterone may also stimulate potassium secretion, but this effect may be counterbalanced by other phenomena and therefore may be dissociated from antinatriuresis.

D. THE ESCAPE PHENOMENON

Very early, investigators reported that the antinatriuretic effect induced by prolonged administration of mineralocorticoids disappeared after a few days of treatment: within 4 to 6 d, sodium excretion and sodium balance returned to control values.[3,66,67] The possible mechanism(s) involved in this escape to the antinatriuretic action of mineralocorticoids were recently reviewed[68-70] and are summarized in Figure 3 and below. Expansion of the vascular compartment brought about by salt and water retention secondary to chronic administration of mineralocorticoids is the starting point of a cascade of events leading to escape. The resulting increased blood pressure enhances sodium excretion through intrarenal mechanisms involved in pressure natriuresis: increased kidney perfusion pressure, decreased release of renin and reduction of angiotensin II level. Extracellular fluid volume expansion also decreases renal sympathic activity,[71] thereby also contributing to decrease renin release.[72] Other mechanisms may be involved in the escape phenomenon: Gauer and Henry[73] first suggested the involvement in the regulation of natriuresis of parameters induced by volume expansion independently of an increase in blood pressure. For example, mineralocorticoid-induced volume expansion triggers, directly as well as through central mechanisms, the synthesis and release of atrial natriuretic peptide (ANP). Prostaglandins synthesized by renal tubules as well as the kinin-kallikrein system also contribute to the escape phenomenon.[69]

ANP inhibits aldosterone production both *in vivo* and *in vitro*,[74,75] is a potent inhibitor of renin release,[76] and consequently of angiotensin II synthesis, thereby magnifying the effect of high renal perfusion pressure on the decrease in angiotensin II level. Furthermore, ANP exerts direct effects on tubular sodium reabsorption, and this raises the controversial question of the site(s) of mineralocorticoid escape along the nephron. Indeed, several studies[49,50] have clearly demonstrated that during escape, the cortical collecting duct, the main renal site of aldosterone action, maintained a high rate of sodium reabsorption as well as potassium secretion. Thus, the nephron segment(s) responsible for the escape might be distinct from the target segment of aldosterone. However, experiments aimed at characterizing these segments led to controversial results: although several studies concluded that during escape, ANP inhibited

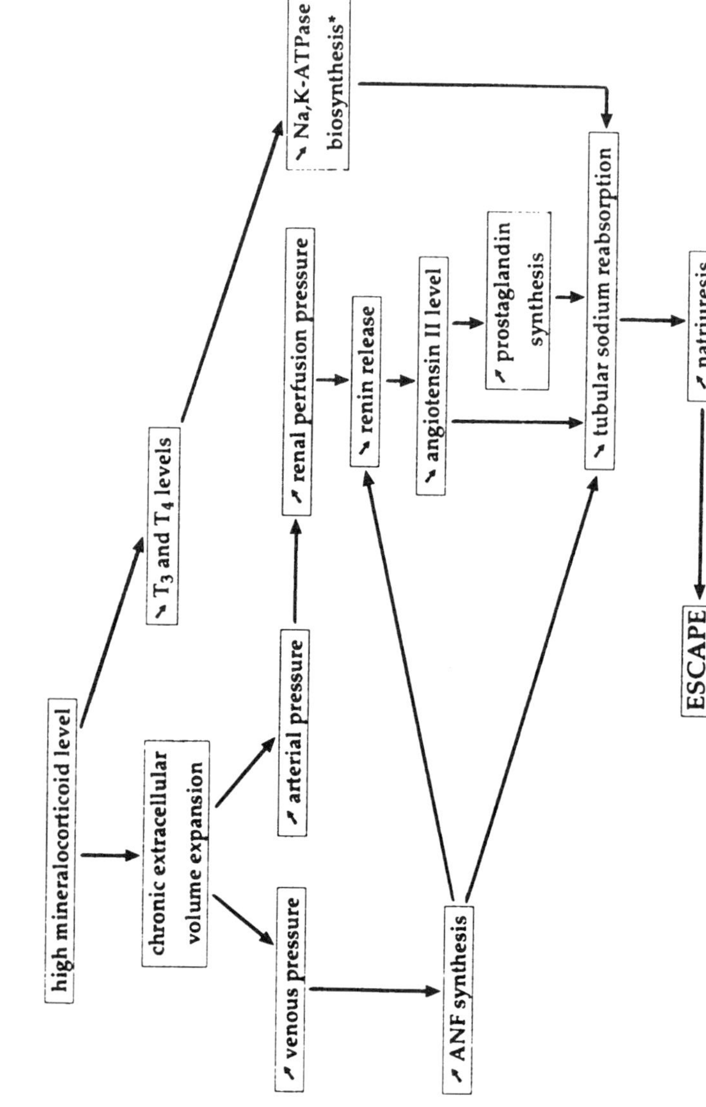

FIGURE 3: Diagram indicating the main mechanisms involved in escape from the sodium-retaining effect of mineralocorticoids. Redrawn with permission from Knox et al.[68]

sodium reabsorption in the proximal tubule,[77,78] Briggs et al.[79] reported that low doses of ANP sufficient to induce natriuresis and diuresis did not alter sodium transport in the proximal tubule and ascending limb of Henle's loop. The inner medullary collecting duct remains a good candidate for escape since it is the main tubular site of action of ANP along the nephron.[80] Prostaglandins, which inhibit sodium reabsorption at different sites along the nephron, including the loop of Henle and the collecting duct,[81] might also play a role.

Recently, Wiener et al.[82] proposed an additional mechanism to account for the escape phenomenon: in rabbits fed a low sodium diet, the resulting increase in plasma aldosterone level was accompanied, after 3 d, with a decrease in triiodothyronine (T_3) plasma level. The latter induced a down-regulation of Na,K-ATPase in the distal colon, which, according to the authors, might participate to the escape phenomenon, if the same were true in the kidney.

In summary, several interrelated mechanisms are involved in the transient character of the antinatriuretic action of aldosterone. It should be stressed that this escape phenomenon takes place along the nephron at sites distinct from the CCD and is restricted to sodium transport, since increased proton and potassium transport persist throughout the duration of mineralocorticoid administration.

E. EFFECT ON URINARY ACIDIFICATION

In the early 1960s, it had already been observed that corticosteroids increased urinary acid excretion,[83,84] but it was not known whether it was a gluco- or a mineralocorticoid effect. In a clearance study, Wilcox et al.[85] demonstrated that acute administration of glucocorticoids or mineralocorticoids to adrenalectomized rats had different effects on urinary acid excretion: dexamethasone promoted net acid output in the urine by increasing the excretion of buffers such as phosphate and ammonia (a function originating mainly in the proximal tubule) whereas aldosterone reduced urine pH.

Stone et al.[86] examined the chronic and acute effects of aldosterone on bicarbonate reabsorption in the rabbit outer medullary collecting duct (OMCD), the main site of urine acidification along the collecting duct due to preponderance of type A intercalated cells. Adrenalectomy reduced the rate of bicarbonate reabsorption by >90%, whereas chronic administration of DOCA almost doubled it. The lumen-positive transepithelial PD prevailing in that nephron segment (due to electrogenic secretion of protons, see Figure 2) varied accordingly. The acute effect of aldosterone was examined by adding the hormone to OMCD harvested from adrenalectomized rabbits and microperfused *in vitro*. Addition of aldosterone 50 n*M* significantly increased bicarbonate reabsorption within 4.5 h; higher concentrations had a more rapid and marked effect (a significant increase was observed as early as 3 h). This action of aldosterone was independent of sodium availability, since it occurred in the presence of amiloride, and curiously did not alter the transepithelial PD. This study clearly demonstrates that the acidifying capacity of the rabbit OMCD is modulated by mineralocorticoids. An acute effect of aldosterone on urinary acidification was also observed by clearance and micropuncture studies in rat inner medullary collecting duct.[87]

F. EFFECT ON WATER TRANSPORT

In addition to electrolytical disorders, Addisonian patients as well as adrenalectomized animals also display a diminished capacity to excrete a water load and to concentrate their urine,[88,89] two defects which may result from the deficiency in glucocorticoids, in mineralocorticoids, or in both.

Now the consensus is that impaired ability to excrete free water is not a direct effect of steroid deficiency in the thick ascending limb of Henle's loop, the site of tubular fluid dilution, but results instead from the decreased delivery of fluid to that segment secondary to decreased glomerular filtration.[90] The concentration defect may be the result of either the low level of antidiuretic hormone (ADH) observed in adrenalectomized animals,[91] the decreased driving force for water reabsorption from the collecting duct brought about by the impaired functioning of the loop of Henle, a decrease in the hydroosmotic response of the collecting duct to ADH, or any combination of the three above. With regard to the sensitivity of the collecting duct to ADH, Schwartz and Kokko[92] reported that water permeability of cortical collecting ducts harvested from adrenalectomized rabbit did not increase in response to ADH, a defect which was corrected within 1.5 hr following *in vitro* addition of either gluco- or mineralocorticoids (at concentrations enabling cross-occupancy of heterologous receptors). Thus, with regard to water handling in the collecting duct, corticosteroids appear to control the regulatory pathway rather than the transport proteins themselves.

IV. MOLECULAR MECHANISM OF ALDOSTERONE ACTION

As previously mentioned, as early as 1963 Edelman et al.[7] had demonstrated that the stimulation of sodium transport observed in response to aldosterone requires RNA and protein synthesis and is dependent on the availability of metabolic substrates. As shown in the model depicted in Figure 4, three distinct molecular targets could account for the effect of aldosterone on sodium transport in the principal cells of the collecting duct: the apical sodium channel, the basolateral Na,K-ATPase, and the mitochondrial enzymes involved in the synthesis of ATP, the fuel required by the pump (Figure 4, top, 1, 2, and 3).

Based on this over-simple model, many studies were undertaken to characterize the proteins which are induced by aldosterone in relation to antinatriuresis and to determine whether their induction is a primary or a secondary effect of the hormone. This analysis is made difficult because of the following: (1) The physiologic effect of a single acute administration of aldosterone appears after a 30 to 60 min latency during which possible feedback regulations may modify the primary hormonal action; (2) In the CCD of mammals, sodium reabsorption is high even in the absence of aldosterone, indicating the presence of numerous transport proteins. Thus, their number is likely, but slightly modified by the hormone, making it difficult to detect these changes; (3) sodium transport is under the control of several interrelated parameters including hormones (glucocorticoids, vasopressin, ANP, and others), local mediators (luminal concentration of sodium, prostaglandins), as well as cellular factors (state of differentiation of the target cell, cell volume, previous stimulation by aldosterone), which may interfere with aldosterone effects or counterbalance them.

A. THE APICAL SODIUM CHANNEL

Apical, amiloride-sensitive sodium channels have been identified in amphibian tight epithelia[28] as well as in the rat cortical collecting duct.[93] Purification of the sodium channel from A6 cells and bovine papillary collecting tubule by Benos et al.[94,95] revealed that the channel molecular weight is 730 kDa and consists of six subunits whose molecular weights vary from 40 to 300 kDa. The amiloride binding site is localized on a 150 kDa subunit and a 95 kDa subunit can be methylated *in vitro*. Very recently, Canessa et al.[96] and Lingueglia et al.[97] have reported independently the expression cloning from rat distal colon of a cDNA encoding a protein of 689 to 699 aminoacids with two putative transmembrane domains. When expressed in Xenopus oocyte, this cDNA encoded a pore-forming protein that was selectively permeant to sodium and was blocked by amiloride. However, the sodium current observed was much smaller than after injecting Xenopus oocytes with size-fractionated mRNAs isolated from rat colon. Along with the properties of the purified channel (see above), this suggests that other subunits may modulate the expression of this cRNA and/or the activity of the pore-forming protein.

Although it is clearly demonstrated that aldosterone increases the apical entry of sodium through these amiloride-sensitive channels,[28] the molecular mechanism of this action remains poorly understood. In the toad urinary bladder, Palmer and colleagues reported that aldosterone increased the density of sodium channels,[98] which may suggest that aldosterone induces their *de novo* synthesis. This hypothesis is supported by several other studies. First, Cuthbert and Shum[99] reported that, in toad bladder, sodium deprivation increased the number of amiloride binding sites in the apical membrane. Secondly, using a specific antibody, Szerlip et al. demonstrated that GP70, a subunit of the amiloride-sensitive sodium-channel, was induced by aldosterone.[100] Other results, however, suggest that the sodium channels are not induced, but would rather be activated by aldosterone. For instance, Garty and Edelman reported that trypsinization of the apical sodium channels of toad bladders prior to aldosterone addition markedly reduced the aldosterone effect on sodium transport,[101] suggesting that the channels which were stimulated by aldosterone pre-existed in the membrane. Also, Kleyman et al. reported that aldosterone altered neither the binding of benzamil (a sodium channel selective-derivative of amiloride)[102] nor the rate of synthesis and of expression of sodium channels at the apical cell surface in A6 cells.[103] Finally, Palmer et al. reported that the level of expression of sodium channels in Xenopus oocytes injected with mRNAs from A6 cells was not affected by pretreatment of these cells with aldosterone.[104]

The apparent discrepancy between the results mentioned above may be explained by the structure of the channel which, as already indicated, consists

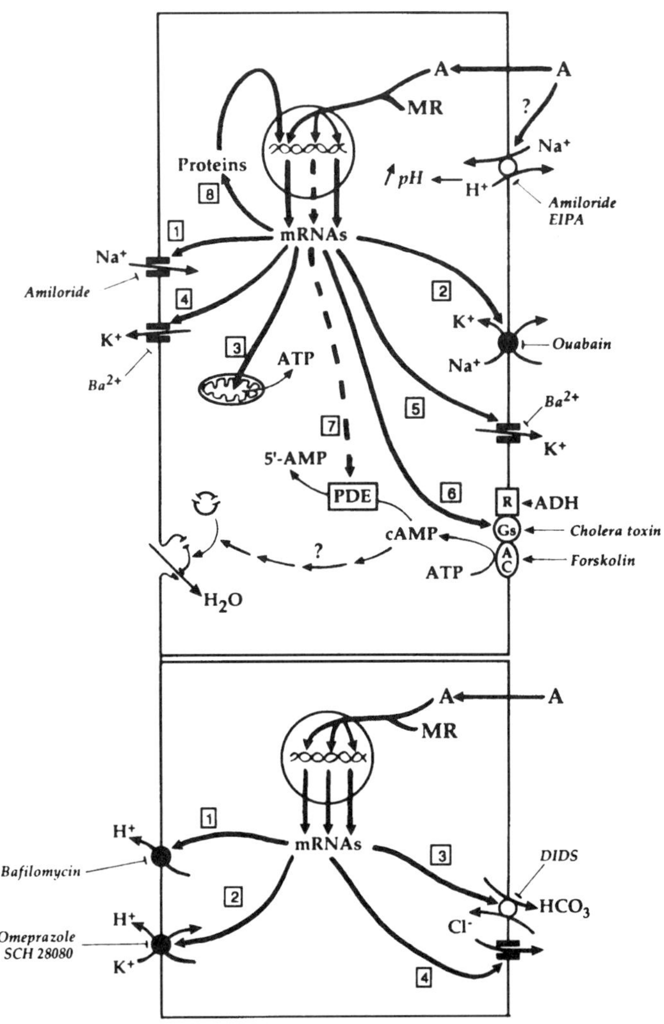
A
A
MR
?
Na+
↑pH
H+
Amiloride
EIPA
Proteins
mRNAs
Na+
Amiloride
1
8
4
K+
Ba2+
3
ATP
2
K+
Ouabain
Na+
Ba2+
K+
5
7
5'-AMP
PDE
6
R
ADH
cAMP
Gs
Cholera toxin
?
A
C
Forskolin
ATP
H2O
A
A
MR
mRNAs
H+
1
Bafilomycin
3
DIDS
2
HCO3
Cl-
H+
Omeprazole
SCH 28080
K+
4

of several subunits, one of which (the one which has been cloned) is a pore-forming unit that binds amiloride. If the other subunits control the activity of the pore-forming one (as suggested by the low sodium current that the latter carries when expressed alone in Xenopus oocyte), it becomes possible to account for the fact that the channels pre-exist in the apical membrane (at least their pore-forming subunits) and that amiloride binding is not modified by aldosterone, whose action would consist of stimulating the biosynthesis of the other subunits of the sodium channel.

Alternatively, aldosterone might activate silent channels pre-existing in the apical membrane through conformational alterations brought about by covalent changes (phosphorylations, methylations, ...) secondary to transient alterations of either intracellular pH or lipidic or oxydative metabolism (reviewed in Reference 42). In this regard, it is noteworthy that the cloned pore-forming protein displays several putative phosphorylation sites. In any case, however, this activation does not influence the probability of channel opening.[105]

B. THE BASOLATERAL NA,K-ATPASE

Na,K-ATPase is a ubiquitous protein that couples the hydrolysis of the energy-rich γ ester-phosphate bound of ATP to the active transport of three sodium ions in exchange with two potassium ions. It thereby generates and maintains the gradients of these cations across cell membranes and contributes to the regulation of many cell functions. Na,K-ATPase consists of a catalytic

---

FIGURE 4: (facing page)   Possible actions of aldosterone in principal cells (top) and type A intercalated cells (bottom) of the collecting duct. In both cell types, aldosterone (A) is supposed to freely enter the cell by diffusion from the bloodstream across the basolateral membrane, to bind to mineralocorticoid receptors (MR) and to migrate within the nucleus where the resulting complex controls the expression of specific genes. In turn, this increases (thick arrows) or decreases (broken arrows) the amount of mRNAs and of the specific proteins they encoded, which accounts for the physiological effects of aldosterone. *Principal cells:* The antinatriuretic effect of aldosterone may result from the stimulation of apical sodium entry [1] by amiloride-inhibitable Na^+ channels, the stimulation of basolateral Na,K-ATPase [2], or increased ATP delivery to the pump through stimulation of mitochondrial metabolism [3]. The kaliuretic effect of aldosterone is in part accounted for by the above with intracellular accumulation of K^+ due to Na,K-ATPase stimulation and depolarization of apical membrane secondary to increased Na^+ entry as consequences; in addition, there might be stimulation of apical [4] and basolateral K^+-channels [5]. Through the latter, K^+ may enter (and not exit) aldosterone-stimulated cells (see the text). The facilitating action of corticosteroids on ADH hydroosmotic effect lies in stimulation of the G_s protein coupling antidiuretic hormone (ADH) receptors to adenylate cyclase (AC) [6] and repression of phosphodiesterase (PDE) which hydrolyzes cyclic AMP (cAMP) into 5′-AMP [7]. Aldosterone-induced mRNAs may also encode for proteins acting as transcription modulators [8]. Finally, increased intracellular pH brought about by direct stimulation of the basolateral amiloride-sensitive Na/H antiporter may interfere with the above mentioned pathways. *Intercalated cells:* Aldosterone induced proteins include the apical bafilomycin-sensitive proton ATPase [1] and the SCH 28080-sensitive H,K-ATPase [1]. Basolateral Cl/HCO_3 antiporter [3] and chloride channels [4] are also possible targets of aldosterone.

α subunit (responsible for hydrolysis of ATP, translocation of cations and ouabain binding), and a β subunit involved in the targeting of *de novo* synthesized Na,K-ATPase units in the plasma membrane and in their stabilization. There are at least three distinct isoforms of each subunit (α_1-α_3, and β_1-β_3) with specific tissue distribution. The mammalian kidney expresses mainly, if not exclusively, the α_1 and β_1 isoforms (reviewed in Reference 106). These isoforms, which are the products of distinct genes, can be in part distinguished functionally: The α_1 isoform is less sensitive to the specific Na,K-ATPase inhibitor ouabain than the α_2 and α_3 isoforms. Na,K-ATPase is present along the whole nephron but its activity is closely related to the sodium transport capacity of the successive segments: in particular, it is quite low in the successive subsegments of the collecting duct.[107]

Several studies have established the relationship that exists between plasma aldosterone level and Na,K-ATPase activity in the collecting duct of mammals. Thus, adrenalectomy leads to reduced Na,K-ATPase activity by over 70% along the collecting duct of rats, rabbits and mice.[61,108,109] Following adrenalectomy, the decrease in Na,K-ATPase occurs after a latency of 1 to 1.5 d, during which a latent pool of Na,K-ATPase is recruited,[110,111] and reaches its maximum after 4 to 5 d,[108] i.e., long after aldosterone has disappeared from plasma in agreement with the half-life of Na,K-ATPase.[112] A single injection of a small dose of aldosterone (2 to 10 µg/kg b.w.) to adrenalectomized rats or rabbits fully restores within 3 h Na,K-ATPase activity up to the level observed in the collecting duct of adrenal-intact animals,[61,109] and this effect is abolished by the antimineralocorticoid spironolactone.[109,113] In adrenal intact animals, however, the time-course of the stimulation of Na,K-ATPase by mineralocorticoids is quite different. Thus, chronic administration of mineralocorticoids to adrenal intact rabbits stimulates Na,K-ATPase activity in their collecting duct but the time of onset and the maximal effects are observed only after 1 to 2 d and 1 week, respectively.[114] In a very elegant study, Hayhurst and O'Neil reported that the latency necessary to detect the stimulation of collecting duct Na,K-ATPase in response to aldosterone was directly related to the initial ATPase activity (and therefore to the initial aldosterone plasma level): the lower the initial Na,K-ATPase activity, the shorter latency period.[115] These findings suggest that additional parameters may modulate aldosterone action on Na,K-ATPase.

One of these parameters has been characterized by studying the *in vitro* action of aldosterone on collecting duct Na,K-ATPase. Indeed, when collecting ducts from adrenalectomized rat were incubated in the presence of aldosterone, a stimulation of Na,K-ATPase remained undetected for 3 h unless triiodothyronine (T_3) was added together with aldosterone.[116] Stimulation was time-dependent (30 min latency, maximal effect within 1.5 to 2 h) and dose-dependent with regard to the two hormones (app. $K_{0.5} \approx 10^{-9}\,M$ and $10^{-10}\,M$ for aldosterone and T_3, respectively). Interestingly, the aldosterone plus T_3 combination also induced a rapid stimulation of Na,K-ATPase activity in the CCD of adrenal intact rats, contrasting with what was observed after *in vivo* administration of aldosterone (the latency being 1 to 2 d). Thus, T_3 might modulate the latency of aldosterone action on Na,K-ATPase, especially since plasma aldosterone and T_3 concentrations are interrelated.[82,117] It is noteworthy that this effect of T_3 is specific for mammals since in amphibian epithelia, aldosterone stimulates sodium transport and Na,K-ATPase in the absence of T_3.[4,42] In fact, in amphibians T_3 has been reported to antagonize the effect of aldosterone.[118]

The last questions to be addressed are: (1) Is Na,K-ATPase activated or induced by aldosterone? (2) If induced, is it induced primarily or secondarily to increased apical sodium entry? These have been long-debated questions which have partially been answered in recent years.

With regard to the first question, it has been shown in the mammalian collecting duct that: (1) stimulation of Na,K-ATPase activity is accompanied with a parallel increase in the number of pump units, as measured by ^{3}H-ouabain binding;[116,119] (2) *in vitro* stimulation of Na,K-ATPase activity is abolished by inhibitors of transcription and of translation;[116] (3) adrenalectomy reduces the amount of mRNAs encoding for the α_1 subunit (but not the β_1 subunit) of Na,K-ATPase, as demonstrated by *in situ* hybridization.[120] This question has received further elements of response from studies in amphibian epithelia or derived cell cultures. In A6 cells, aldosterone increases very early the transcription rate of the mRNAs encoding for both Na,K-ATPase subunits. This effect is observed as early as 15 to 30 min after addition of aldosterone,[121] and within 6 h the amount of mRNAs is increased four- and twofold for the β_1 and the α_1 subunits, respectively.[122] Although aldosterone rapidly induces the transcription of Na,K-ATPase genes, the expression of the newly synthesized pump units in the basolateral membrane is only observed after 6 to 18 h of latency.[118,123,124] This delayed expression of Na,K-ATPase in the basolateral membrane of amphibian epithelial cells is accounted for by the time required for accumulation of α_1 and β_1 mRNAs, translation into proteins, and assembly of α β subunits and insertion into the membrane.[125,126] That this process takes much longer in amphibian than in mammalian cells may be related to the difference between the effects of T_3 in these two groups of animals (see above). It is also noteworthy that the metabolism is much slower in poikilotherms than in homeotherms.

These results also clearly demonstrate that stimulation of the transcription of Na,K-ATPase subunit genes is a primary effect of aldosterone in the amphibian epithelia. In the mammalian collecting duct, this conclusion has been controversial for a long time, as it was assumed that aldosterone-induced stimulation of Na,K-ATPase in the CCD was secondary to increased apical sodium entry, because *in vivo* pretreatment of adrenalectomized rabbit with amiloride prior to aldosterone injection abolished Na,K-ATPase stimulation.[127] However, when re-evaluated *in vitro*, it was shown that at a concentration sufficient to block totally and specifically sodium channels,[93] amiloride did not alter the induction of Na,K-ATPase by aldosterone in the CCD,[116] demonstrating that it is independent from an increment of intracellular concentration of sodium brought about by increased apical sodium conductance. In fact, the inhibitory effect of amiloride observed *in vivo* might be due to inhibition of the Na/H antiporters rather than of the sodium channels. Indeed, Oberleithner et al. reported that aldosterone increases intracellular pH through activation of the Na/H exchanger in cells of the distal nephron of the frog.[128] Furthermore, specific inhibition of the Na/H exchanger by ethylisopropylamiloride (EIPA) abolishes the *in vitro* induction of Na,K-ATPase in the rat collecting duct which can be antagonized by alkalinizing the cells by incubation at higher pH.[129]

In summary, these studies demonstrate that in the mammalian collecting duct Na,K-ATPase is a molecular target of aldosterone which under certain conditions can be induced very early, with the time-course characterizing sodium transport.[61] However, the latency for the appearance of active pump units in the membrane varies with the initial status of the animals, in particular with their aldosterone and T_3 status. In amphibians where T_3 is not involved in that process, aldosterone induces the transcription of Na,K-ATPase genes very early, but the setting in place of the new pump units is delayed.

C. OXIDATIVE METABOLISM

Because aldosterone-induced antinatriuresis is dependent on cellular energy supply by oxidative metabolism and is blunted under anoxic conditions, it has been postulated that metabolic enzymes, in particular enzymes of the tricarboxylic acid-cycle, might be aldosterone targets.[130] Accordingly, several enzymes of the tricarboxylic acid-cycle and proteins linked to oxidative phosphorylation are up-regulated by aldosterone in toad bladder or mammalian kidney.[131-135] Particular attention has been paid to the regulation of the mitochondrial citrate synthase by aldosterone. Law and Edelman reported that citrate synthase activity decreased in the kidney of adrenalectomized rats and was stimulated as early as 3 h after injection of aldosterone.[136] In toad bladder, aldosterone also stimulates citrate synthase activity, and its effect is blocked by inhibitors of RNA and protein synthesis,[132] suggesting that aldosterone induces citrate synthase synthesis. Marver and Schwartz also reported that in rabbit CCD adrenalectomy decreased by half citrate synthase activity and that administration of aldosterone restored it within 1.5 h.[137] Altogether, these results suggest that citrate synthase is an early aldosterone-induced protein. However, the induction of citrate synthase is not an absolute prerequisite for the overall antinatriuretic effect of aldosterone since it has been reported that aldosterone enhances sodium transport in several cultured cell lines (derived from toad urinary bladder or kidney), without change in the activity of citrate synthase.[138]

In conclusion, although citrate synthase is likely induced by aldosterone, this action is not an absolute requirement for the development of a full antinatriuretic response. This is in agreement with the fact that other hormones such as ADH may increase sodium transport in CCD without altering citrate synthase activity.[132]

D. POTASSIUM CHANNELS

By itself, the stimulation of apical sodium channels and of basolateral Na,K-ATPase, which supports aldosterone action on transepithelial sodium transport, would be sufficient to account for the hormone effect on potassium transport in the principal cells of CCD. Indeed, as depicted in the cellular model in Figure 2, potassium secretion proceeds through its active pumping into the cell by the basolateral Na,K-ATPase, thereafter, leaking passively through Ba^{2+}-sensitive apical potassium channels. Therefore, stimulation of Na,K-ATPase favors the accumulation of potassium within the cells and stimulation of apical sodium entry depolarizes the apical membrane, thereby increasing the driving force for potassium exit at this border of the cell. Nonetheless, several reports indicate that aldosterone may also control the activity of specific potassium transport systems, in particular the apical and basolateral potassium channels (see Figure 4, top, 4 and 5), which would allow for some dissociation between aldosterone effects on sodium reabsorption and on potassium secretion.

Sansom and O'Neil reported that chronic administration of DOCA to rabbits doubled the Ba^{2+}-sensitive conductance of the apical membrane of CCD cells within 4 d of treatment.[139] In adrenalectomized rabbits, this stimulatory effect of aldosterone occurs as soon as after 3 h.[140] Thus, aldosterone would not only increase the driving force for potassium secretion across the apical membrane, but also the conductance of this membrane. However, this effect of aldosterone on apical potassium conductance is not found in all species. In the rat CCD, for example, Schafer et al. failed to demonstrate any change in potassium conductance of the apical membrane in response to chronic DOCA treatment.[141,142] Similarly, patch-clamp studies indicated that feeding rats with a low-salt diet modified neither the density of apical potassium channels nor their conductance or open probability.[143] Thus, in the rat CCD the main effect of aldosterone on the apical exit of potassium is exclusively on the driving force.

In the rabbit, adrenalectomy decreases the potassium conductance of the basolateral membrane of the CCD[141] whereas chronic DOCA treatment increases it.[145] This latter effect was first thought to be involved in the stimulation of transepithelial sodium reabsorption by facilitating the recycling potassium accumulated within the cell by Na,K-ATPase across the basolateral membrane. However, under conditions of high sodium transport, the basolateral membrane is hyperpolarized so that the prevailing electrochemical gradient for potassium favors its entry into the cells[146] rather than its exit. Thus, coupled to the enhanced conductance for potassium of the apical membrane, changes at the basolateral border of the cell contribute to increased potassium secretion into CCD lumen.

Given the molecular events underlying aldosterone action on cation transport in CCD, it becomes obvious that part of the kaliuretic effect (which is accounted for by apical membrane depolarization and basolateral membrane hyperpolarization) is tightly coupled to the antinatriuretic effect, whereas the part resulting from increased potassium conductance of the two cell membranes might be independent from increased sodium transport.

Based on the observation that in CCD from adrenalectomized rabbits amiloride abolishes the early increase of potassium conductance of the apical membrane promoted by aldosterone, Sansom et al. proposed that this effect was secondary to increased sodium reabsorption.[140] However, as already discussed for Na,K-ATPase, this effect of amiloride is more likely to be secondary to inhibition of the basolateral Na/H exchanger than to inhibition of apical sodium channels. In MDCK cells, a dog kidney-derived cell line exhibiting some properties of collecting duct cells, aldosterone also stimulates the Na/H antiporter and alcalinizes the cells; it thereby increases potassium secretion via the setting of pre-existing potassium channels in the apical membrane.[147] In the frog distal renal tubule, Wang et al. also reported that within 1 d, aldosterone enhanced the density of apical potassium channels.[148] This effect was mimicked by increasing intracellular pH and was markedly reduced both by amiloride and ouabain, which is consistent with the involvement of a stimulation of the Na/H antiporter. Thus, it is most likely that regulation of apical potassium channels by aldosterone results from the recruitment of inactive channels rather that de novo synthesis of new channels. This action is independent of the antinatriuretic action of aldosterone but is secondary to the activation of the Na/H exchanger. Since this is probably not a genomic effect of aldosterone (see Chapter 5), increased apical potassium conductance might be insensitive to actinomycin D.

Finally, it should be kept in mind that the above mentioned effects of aldosterone on potassium secretion by principal cells of the collecting duct might be blunted in part by concomitant stimulation of the apical, potassium-reabsorbing H,K-ATPase present in intercalated cells (see below).

E. PROTON TRANSPORT SYSTEMS

Based on the cellular model generally admitted for proton secretion and bicarbonate reabsorption in type A intercalated cells (see Figure 2), several putative molecular targets could qualify for aldosterone action on urinary acidification: the apical proton translocating ATPases (H-ATPase and H,K-ATPase), the basolateral Cl/HCO_3 exchanger and the basolateral chloride channel (Figure 4, bottom, 1 to 4).

The electrogenic V-type proton ATPase is down-regulated by adrenalectomy in the rat collecting duct[149] and is up-regulated by chronic administration of aldosterone to rats and rabbits.[150,151] Aldosterone also promotes a short-term (1 to 2 h) stimulation of H-ATPase activity when administered to adrenalectomized rats or when added *in vitro* to collecting tubules harvested from adrenalectomized rats.[152] Although this latter effect is abolished by inhibitors of synthesis of mRNAs and of proteins, it is not known whether aldosterone induces the *de novo* synthesis of H-ATPase subunits or of regulatory proteins involved in the insertion of new pumps in the apical membrane by fusion of subapical vesicles, a usual mechanism of regulation of this pump.[153]

In the rabbit, Garg has reported that chronic administration of aldosterone increased H,K-ATPase in the CCD,[154] whereas we have recently observed that adrenalectomy decreased this activity in the rat collecting duct.[155] It is not possible to conclude from results of such *in vivo* manipulations of the mineralocorticoid status whether these changes in H,K-ATPase activity are primary effects of aldosterone or result from changes of other parameters such as acid/base balance or kalemia, which are known to control this pump activity.[156-158] However, stimulation of H,K-ATPase *in vitro* by addition of aldosterone has been reported in MDCK cells,[159] demonstrating a direct control by the hormone.

Thus, it is likely that the two proton pumps present in type A intercalated cells are up-regulated by aldosterone. On the other hand, it is not known whether, in addition, aldosterone influences the basolateral Cl/HCO$_3$ exchanger and/or the chloride channel.

F. ADH SIGNALING PATHWAY

It is well established that the hydroosmotic effect of ADH on the principal cells of the collecting duct is mediated through V$_2$ receptors and the cyclic AMP cascade. The first biochemical evidence that this signaling pathway is altered by corticosteroids was provided by Rajerison et al. who observed that adrenalectomy reduced ADH-sensitive adenylate cyclase (AC) activity and the number of ADH receptors in rat kidney.[160] Because adrenalectomy reduced ADH-sensitive AC activity to a larger extent than the number of ADH receptors and did not alter the catalytic activity of the AC system, these authors concluded that adrenalectomy impaired the efficiency of the coupling between ADH receptors and AC.

Impaired coupling was recently demonstrated at the level of the isolated collecting duct. In this nephron segment, adrenalectomy reduced to the same degree ADH-sensitive as well as cholera-toxin stimulated AC activity, whereas it did not modify forskolin-stimulated AC.[161] These findings demonstrate that adrenalectomy alters the Gs protein which couples the V$_2$ receptor of ADH to AC (Gs protein which can be activated by cholera toxin independently of receptor occupancy), but not the catalytic activity of AC (which can be stimulated directly by forskolin) (Figure 4, 6). This effect is entirely accounted for by mineralocorticoid deficiency since it is reversed within 2.5 h *in vitro* by aldosterone. Furthermore, the *in vitro* effect of aldosterone is blocked by the antimineralocorticoid spironolactone but not by RU 38486, a specific glucocorticoid antagonist.[162] Finally, this short-term effect of aldosterone is abolished by inhibitors of mRNAs and protein synthesis.

Adrenalectomy also impairs the hydroosmotic effect of ADH in the inner medullary collecting duct, but the molecular mechanism is different. Indeed, Jackson et al. have demonstrated that adrenalectomy is associated with increased phosphodiesterase activity (Figure 4, 7), thereby lessening the rise in intracellular cAMP level induced by ADH.[163] However, this effect appears as secondary to gluco- rather than mineralocorticoid deficiency.

V. CONCLUSION

In this chapter, we attempted to review most of the biochemical and physiological studies from which our present understanding of the genomic effects of aldosterone on the kidney has emerged. Multiple experimental approaches led to clear identification of several proteins (either ion transporters or regulatory proteins) the activity of which is controlled by aldosterone. Except for a few among them (such as α and β subunits of Na,K-ATPase), it is not clearly established as yet whether mineralocorticoids modulate expression or activity of those proteins. In a near future, molecular cloning of the promoter of the genes encoding for these proteins will provide a final answer to these questions, since the DNA sequences recognized by the mineralocorticoid receptor are well characterized. At that time, two key questions will remain. First, it will be necessary to identify the transcription factors induced in the early phase of aldosterone action (Figure 4, top, 8). Second, it will be of major importance to characterize the post-transcriptional regulation of induced proteins because these steps determine the overall physiologic action of aldosterone. These regulations may be accounted for by genomic as well as non-genomic actions of aldosterone (for example, intracellular alkalinization) and also by interaction with other unidentified parameters.

ACKNOWLEDGMENTS

We are indebted to V. Biausque and L. Bonnet-Lericque for their help during preparation of this manuscript. Research from our laboratory discussed in this review was partly supported by the Centre National de la Recherche Scientifique.

REFERENCES

1. **Luetscher, J. A., Jr.,** Studies of aldosterone in relation to water and electrolyte balance in man, *Recent Prog. Horm. Res.,* 12, 175, 1956.
2. **Barger, A. C., Berlin, R. D., Tulenko, J. F.,** Infusion of aldosterone 9-α-fluorohydrocortisone and antidiuretic hormone into the renal artery of normal and adrenalectomized, unanesthetized dogs: Effect on electrolyte and water excretion, *Endocrinology,* 62, 804, 1958.
3. **August, C., Nelson, D. Thorn, G. W.,** Response of normal subjects to large amounts of aldosterone, *J. Clin. Invest.,* 37, 1549, 1958.
4. **Crabbé, J.,** Stimulation of active sodium transport by the isolated toad bladder with aldosterone in vitro, *J. Clin. Invest.,* 40, 2103, 1961.

5. **Sharp, G. W., Leaf, A.,** Mechanism of action of aldosterone, *Phys. Rev.,* 46, 593, 1966.
6. **Vander, A. J., Malvin, R. L., Wilde, W. S.,** Effects of adrenalectomy and aldosterone on proximal and distal tubular sodium reabsorption, *Proc. Soc. Exp. Biol. Med.,* 99, 323, 1958.
7. **Edelman, I. S., Bogoroch, R., Porter, G. H.,** On the mechanism of action of aldosterone on sodium transport: the role of protein synthesis, *Proc. Natl. Acad. Sci. U.S.A.,* 50, 1169, 1963.
8. **Edelman, I. S.,** Mechanism of action of steroid hormones, *J. Steroid Biochem.,* 6, 147, 1975.
9. **Rossier, B. C.,** Mechanisms of action of mineralocorticoid hormones, *Endocr. Res.,* 15, 203, 1989.
10. **Herman, K., Nawaka, H., Kato, K., Ibayashi, H., Matsuo, H.,** Studies on renal aldosterone-binding proteins, *J. Biol. Chem.,* 243, 3849, 1968.
11. **Evans, R. M.,** The steroid and thyroid hormone receptor superfamily, *Science,* 240, 889, 1988.
12. **Härd, T., Kellenbach, E., Boelens, R., Maler, B. A., Dahlman, K., Freedman, L. P., Carlstedt-Duke, J., Yamamoto, K. R., Gustaffson, J., Kaptein, R.,** Solution structure of the glucocorticoid receptor DNA-binding domain, *Science,* 249, 157, 1990.
13. **Beato, M.,** Gene regulation by steroid hormones, *Cell,* 89, 335, 1989.
14. **Green, S., Chambon, P.,** Nuclear receptors enhance our understanding of transcription regulation, *Trends Genet.,* 4, 309, 1988.
15. **Fawell, S. E., Lees, J. A., White, R., Parker, M. G.,** Characterization and colocalization of steroid binding and dimerization activities in the mouse estrogen receptor, *Cell,* 60, 953, 1990.
16. **Picard, D., Kumar, V., Chambon, P., Yamamoto, K. R.,** Signal transduction by steroid hormones: Nuclear localization is differentialy regulated in estrogen and glucocorticoid receptors, *Cell Regul.,* 1, 291, 1990.
17. **Picard, D., Dhursheed, B., Garabedian, M. J., Fortin, M. G., Lindquist, S., Yamamoto, K. R.,** Signal transduction by steroid receptors: Reduced levels of HSP90 compromise receptor action in vitro, *Nature,* 348, 166, 1990.
18. **Arriza, R. M., Weinberger, C., Cerelli, G., Glaser, T. M., Handelin, B. L., Housman, D. E., Evans, R. M.,** Cloning of human mineralocorticoid receptor complementary cDNA: Structural and functional kinship with the glucocorticoid receptor, *Science,* 237, 268, 1987.
19. **Rafestin-Oblin, M. E., Couette, B., Radanyi, C., Lombès, M., Baulieu, E. E.,** Mineralocorticosteroid receptor of the chick intestine: Oligomeric structure and transformation, *J. Biol. Chem.,* 264, 9304, 1989.
20. **Alnemri, E. S., Maksymowych, A. B., Robertson, N. M., Litwack, G.,** Overexpression and characterization of the human mineralocorticoid receptor, *J. Biol. Chem.,* 266, 18072, 1991.
21. **Schulman, G., Bodine, P. V., Litwack, G.,** Modulators of the glucocorticoid receptor also regulate mineralocorticoid receptor function, *Biochemistry,* 31, 1734, 1992.
22. **Arriza, J. L., Simerly, R. B., Swanson, L. W., Evans, R. M.,** The neuronal mineralocorticoid receptor as a mediator of glucocorticoid response, *Neuron,* 1, 887, 1988.
23. **Pearce, D., Yamamoto, K. R.,** Mineralocorticoid and glucocorticoid receptor activities distinguished by non receptor factors at a composite response element, *Science,* 259, 1161, 1993.
24. **Diamond, M. I., Miner, J. N., Yoshinaga, S. K., Yamamoto, K. R.,** Transcription factor interactions: Selectors of positive or negative regulation from a single DNA element, *Science,* 249, 1266, 1990.
25. **Zhang, X., Dong, J., Chiu, J.,** Regulation of α–fetoprotein gene expression by antagonism between AP-1 and the glucocorticoid receptor at their overlapping binding site, *J. Biol. Chem.,* 266, 8248, 1991.
26. **Funder, J. W.,** Mineralocorticoids, glucocorticoids, receptors and response elements, *Science,* 259, 1132, 1993.

27. **Doucet, A., Barlet-Bas, C.,** Involvement of Na$^+$,K$^+$-ATPase in antinatriuretic action of mineralocorticoids in mammalian kidney, *Curr. Top. Membr. Transp.,* 34, 185, 1989.

28. **Garty, H.,** Mechanisms of aldosterone action in tight epithelia, *J. Membr. Biol.,* 90, 193, 1986.

29. **Kriz, W., Bankir, L.,** A standard nomenclature for structures of the kidney, *Kidney Int.,* 33, 1, 1988.

30. **Morel, F., Doucet, A.,** Functional segmentation of the nephron, in *The Kidney: Physiology and Physiopathology,* Seldin, D.W., Giebisch, G., Eds., Raven Press, New York, 1992, 1049.

31. **Burg, M. B., Grantham, J., Abramow, M., Orloff, J.,** Preparation and study of fragments of single rabbit nephrons, *Am. J. Physiol.,* 210, 1293, 1966.

32. **Frömter, E.,** Solute transport across epithelia: what we can learn from micropuncture studies on kidney tubules, *J. Physiol.,* 288, 1, 1979.

33. **Doucet, A.,** Multiple hormonal control of kidney tubular functions, *News in Physiological Sciences,* 2, 141, 1987.

34. **Ussing, H. H., Zerahn, K.,** Active transport of sodium as the source of electric current in the short-circuited isolated frog skin, *Acta Physiol. Scand.,* 23, 100, 1951.

35. **Koeffoed-Johnson, V., Ussing, H. H.,** The nature of frog skin potential, *Acta Physiol. Scand.,* 43, 298, 1958.

36. **Rafferty, K. A., Jr.,** Mass culture of amphibian cells: Methods and observations concerning stability of cell type, in *Biology of Amphibian Tumors,* Mizell., M., Eds., Springer-Verlag, New-York, 1969, 52.

37. **Perkins, F. M., Handler, J. S.,** Transport properties of toad kidney epithelia in culture, *Am. J. Physiol.,* 241, C154, 1981.

38. **Spooner, P. M., Edelman, I. S.,** Further studies on the effect of aldosterone on electrical resistance of toad bladder, *Biochim. Biophys. Acta,* 406, 304, 1975.

39. **Lichtenstein, N. S., Leaf, A.,** Effect of amphotericin B on the permeability of the toad bladder, *J. Clin. Invest.,* 44, 1328, 1965.

40. **Crabbé, J.,** Suppression by amphotericin B of the effect exerted by aldosterone on active sodium transport, *Arch. Int. Physiol. Biochim.,* 75, 342, 1967.

41. **Nagel, W., Crabbé, J.,** Mechanism of action of aldosterone on active sodium transport across toad skin, *Pflügers Arch.,* 385, 181, 1980.

42. **Rossier, B. C., Palmer, L.,** Mechanisms of aldosterone action on sodium and potassium transport, in *The Kidney: Physiology and Physiopathology,* Seldin, D. W., Giebisch, G., Eds., Raven Press, New York, 1991, 1373.

43. **Horisberger, J. D., Diezi, J.,** Effects of mineralocorticoids on Na$^+$ and K$^+$ excretion in the adrenalectomized rat, *Am. J. Physiol.,* 254, F89, 1983.

44. **Campen, T. J., Vaughn, D. A., Fanestil, D. D.,** Minerelo- and glucocorticoid effects on renal excretion of electrolytes, *Pflügers Arch.,* 399, 93, 1983.

45. **Williamson, H. E.,** Mechanism of the antinatriuretic action of aldosterone, Biochem. Pharmacol., 12, 1449, 1963.

46. **Fimognari, G. M., Fanestil, D. D., Edelman, I. S.,** Induction of RNA and protein synthesis in the action of aldosterone in the rat, *Am. J. Physiol.,* 213, 954, 1967.

47. **Horisberger, J. D., Diezi, J.,** Inhibition of aldosterone-induced antinatriuresis and kaliuresis by actinomycin D, *Am. J. Physiol.,* 246, F101, 1984.

48. **Gross, J. B., Imai, M., Kokko, J. P.,** A functional comparison of the cortical collecting tubule and the distal convoluted tubule, *J. Clin. Invest.,* 55, 1284, 1975.

49. **O'Neil, R. G., Helman, S. I.,** Transport characteristics of renal collecting tubules: Influences of DOCA and diet, *Am. J. Physiol.,* 232, F544, 1977.

50. **Schwartz, G. J., Burg, M. B.,** Mineralocorticoid effects on cation transport by cortical collecting tubules *in vitro, Am. J. Physiol.,* 253, 576, 1978.

51. **Stokes, J. B., Ingram, M. J., Williams, A. D., Ingram, D.,** Heterogeneity of the rabbit collecting tubule: localization of mineralocorticoid hormone action to the cortical portion, *Kidney Int.,* 20, 340, 1981.

52. **Imai, M.,** The connecting tubule: a functional subdivision of the rabbit distal segments, *Kidney Int.,* 15, 346, 1979.

53. **Wade, J. P., O'Neil, R. G., Pryor, J. L., Boulpaep, E. L.,** Modulation of cell membrane area in renal collecting tubules by corticosteroid hormones, *J. Cell Biol.,* 81, 439, 1979.

54. **Gross, J. B., Kokko, J. P.,** Effects of aldosterone and potassium sparing diuretics on electrical potential difference across the distal nephron, *J. Clin. Invest.,* 59, 82, 1977.

55. **Wingo, C. S., Kokko, J. P., Jacobson, H. R.,** Effects of *in vitro* aldosterone on the rabbit cortical collecting tubule, *Kidney Int.,* 28, 51, 1985.

56. **Allen, C. G., Barrat, L. J.,** Effect of aldosterone on the transepithelial potential difference of the rat tubule, *Kidney Int.,* 19, 678, 1981.

57. **Husted, R. F., Laplace, J. R., Stokes, J. B.,** Enhancement of electrogenic Na$^+$ transport across rat inner medullary collecting duct by glucocorticoid and mineralocorticoid hormones, *J. Clin. Invest.,* 86, 498, 1990.

58. **Rabinovitz, L.,** Aldosterone and renal potassium excretion, *Renal Physiol.,* 2, 229, 1979.

59. **Stokes, J. B.,** Potassium secretion by cortical collecting tubule: relation to sodium absorption, luminal Na$^+$ concentration, and transepithelial voltage, *Am. J. Physiol.,* 241, F395, 1981.

60. **Bia, M. F., Karen, T., DeFronzo, R. A.,** The effect of dexamethasone on renal electrolyte excretion in the adrenalectomized rat, *Endocrinology,* 111, 882, 1982.

61. **El Mernissi, G., Doucet, A.,** Short-term effect of aldosterone on renal sodium transport and tubular Na-K-ATPase in the rat, *Pflügers Arch.,* 399, 139, 1983.

62. **Lifschitz, M. D., Schrier, R. W., Edelman, I. S.,** Effect of actinomycin D on aldosterone mediated changes in electrolyte excretion, *Am. J. Physiol.,* 224, 376, 1973.

63. **Wiederholt, M., Behn, C., Schoormans, W., Hansen, L.,** Effect of aldosterone on sodium and potassium transport in the kidney, *J. Steroid Biochem.,* 3, 151, 1972.

64. **Howell, D. S., Davis, J. O.,** Relationship of sodium retention to potassium excretion by the kidney during administration of desoxycorticosterone acetate to dogs, *Am. J. Physiol.,* 179, 359, 1954.

65. **Peterson, L. N., Wright, F. S.,** Effect of sodium intake on renal potassium excretion, *Am. J. Physiol.,* 233, F225, 1977.

66. **Daughaday, W. H., Mabryde, C. M.,** Renal and adrenal mechanisms of salt conservation: The excretion of urinary formaldehydogenic steroids and 17-ketosteroids during salt deprivation and desoxycorticosterone administration, *J. Clin. Invest.,* 29, 591, 1950.

67. **Relman, A. S., Schwartz, W. B.,** The effect of DOCA on electrolyte balance in normal man and its relation to sodium chloride intake, *Yale J. Biol. Med.,* 24, 540, 1952.

68. **Knox, F. G., Burnett, J. C., Jr., Kohan, D. E., Spielman, W. S., Strand, J. C.,** Escape from the sodium-retaining effects of mineralocorticoids, *Kidney Int.,* 17, 263, 1980.

69. **Gonzalez-Campoy, J. M., Romero, J. C., Knox, F. G.,** Escape from the sodium-retaining effects of mineralocorticoids: Role of ANF and intrarenal hormone systems, *Kidney Int.,* 35, 767, 1989.

70. **Haas, J. A., Knox, F. G.,** Mechanisms for escape from the salt-retaining effects of mineralocorticoids: role of deep nephrons, *Semin. Nephrol.,* 10, 380, 1990.

71. **Romero, J. C., Knox, F. G., Opgenorth, T. J., Granger, J. P., Keiser, J. A.,** Contribution of sympathetic neural reflexes to mineralocorticoid escape, *Fed. Proc.,* 44, 2382, 1985.

72. **Blair, M. L., Chen, Y. H., Hisa, H.,** Elevation of plasma renin activity by alpha-adrenoreceptor agonists in conscious dogs, *Am. J. Physiol.,* 251, E695, 1896.

73. **Gauer, D. H., Henry, J. P.,** Neurohormonal control of plasma volume, in *International Review of Physiology, Cardiovascular Physiology II,* Vol. 9, Guyton A., Cowley, A., Eds., University Park Press, Baltimore, 1976, 145.

74. **Chartier, L., Schiffrin, E. L., Thibault, G., Garcia, R.,** Atrial natriuretic factor inhibits the effects of angiotensin II, ACTH and potassium on aldosterone secretion *in vitro* and angiotensin II-induced steroidogenesis *in vivo, Endocrinology,* 115, 2026, 1984.

75. **Chartier, L., Schiffrin, E. L.,** Atrial natriuretic peptide inhibits the effect of endogenous angiotensin II on plasma aldosterone in conscious sodium-depleted rats, *Clin. Sci.,* 72, 31, 1987.

76. **Burnett, J. C., Jr., Granger, J. P., Opgenorth, T. J.,** Effects of synthetic atrial natriuretic factor on renal function and renin release, *Am. J. Physiol.,* 247, F863, 1984.

77. **Hammond, T., Yusufi, A. N. K., Knox, F. G.,** Administration of atrial natriuretic factor inhibits sodium-coupled transport in proximal tubules, *J. Clin. Invest.,* 75, 1983, 1985.

78. **Biollaz, J., Bidiville, J., Diezi, J., Waeber, B., Nussberger, J., Brunner-Ferber, F., Gomez, H. J., Brunner, H. R.,** Site of action of a synthetic atrial natriuretic peptide evaluated in humans, *Kidney Int.,* 32, 537, 1987.

79. **Briggs, J. P., Steipe, B., Schubert, G., Schermann, J.,** Micropuncture studies of renal effects of atrial natriuretic substances, *Pflügers Arch.,* 395, 271, 1982.

80. **Nonoguchi, H., Sands, J. M., Knepper, M. A.,** Atrial natriuretic factor inhibits vasopressin-stimulated osmotic water permeability in rat inner medullary collecting duct, *J. Clin. Invest.,* 82, 1383, 1988.

81. **Bonvalet, J. P., Pradelles, P., Farman, N.,** Segmental synthesis and actions of prostaglandins along the nephron, *Am. J. Physiol.,* 253, F377, 1987.

82. **Wiener, H., Nielsen, J. M., Klaerke, D. A., Jorgensen, P. L.,** Aldosterone and thyroid hormone modulation of α_1-, β_1-mRNA, and Na,K-pump sites in rabbit distal colon epithelium. Evidence for a novel mechanism of escape from the effect of hyperaldosteronemia, *J. Membr. Biol.,* 133, 203, 1993.

83. **Bartter, F. C., Fourman, P.,** The different effects of aldosterone-like steroids and hydrocortisone-like steroids on urinary excretion of potassium and acid, *Metabolism,* 11, 6, 1962.

84. **Mills, J. N., Thomas, S., Williamson, K. S.,** The acute effects of hydrocortisone, deoxycorticosterone and aldosterone upon the excretion of Na, K, and acid by the human kidney, *J. Physiol. (London),* 151, 312, 1960.

85. **Wilcox, C. S., Cemerikic, D. A., Giebisch, G.,** Differentiel effects of acute mineralo- and glucocorticosteroid administration on renal acid elimination, *Kidney Int.,* 21, 546, 1982.

86. **Stone, D. K., Seldin, D. W., Kokko, J. P., Jacobson, H. R.,** Mineralocorticoid modulation of rabbit medullary collecting duct acidification, *J. Clin. Invest.,* 72, 77, 1983.

87. **Higashihara, E., Carter N. W., Pacacco, L., Kokko, J. P.,** Aldosterone effects on papillary collecting duct pH profile of the rat, *Am. J. Physiol.,* 246, F726, 1984.

88. **Willson, D. M., Sunderman, F. W.,** Studies in serum electrolytes. XII. The effect of water restriction in patients with Addison's disease receiving sodium chloride, *J. Clin. Invest.,* 18, 35, 1939.

89. **Green, H. H., Harrington, A. R., Valtin, H.,** On the role of antidiuretic hormone in the inhibition of acute water diuresis in adrenal insufficiency and the effects of gluco- and mineralocorticoids in reversing the inhibition, *J. Clin. Invest.,* 49, 1724, 1970.

90. **Schrier, R. W., Bichet, D. G.,** Osmotic and non osmotic control of ADH release and the pathogenesis of impaired water excretion in adrenal, thyroid and edematous disorders, *J. Lab. Clin. Med.,* 98, 1, 1981.

91. **Seif, S. M., Robinson, A. G., Zimmerman, E. A., Wilins, J.,** Plasma neurophysin and ADH in the rat: response to adrenalectomy and steroid replacement, *Endocrinology,* 103, 1009, 1978.

92. **Schwartz, M. J., Kokko, J. P.,** Urinary concentrating defect of adrenal insufficiency. Permissive role of adrenal steroids on the hydroosmotic response across the rabbit cortical collecting tubule, *J. Clin. Invest.,* 66, 234, 1980.

93. **Palmer, L. G., Frindt, G.,** Amiloride-sensitive Na channels from the apical membrane of the rat cortical collecting tubule, *Proc. Natl. Acad. Sci. U.S.A.,* 83, 2767, 1986.

94. **Sariban-Sohraby, S., Benos, D. J.,** Detergent solubilization, functional reconstitution, and partial purification of epithelial amiloride-binding protein, *Biochemistry,* 25, 4639, 1986.

95. **Benos, D. J., Saccomani, G., Brenner, B. M., Sariban-Sohraby, S.,** Purification and characterization of the amiloride-sensitive sodium channel from A6 cultured cells and bovine renal papilla, *Proc. Natl. Acad. Sci. U.S.A.,* 83, 8525, 1986.

96. **Canessa, C. M., Horisberger, J.-D., Rossier, B. C.,** Epithelial sodium channel related to proteins involved in neurodegeneration, *Nature,* 361, 467, 1993.

97. **Lingueglia, E., Voilley, N., Waldmann, R., Lazdunski, M., Barbry, P.,** Expression cloning of an epithelial amiloride-sensitive Na^+ channel. A new channel type with homologies to *C. elegans* degeneresis, *FEBS Lett.,* 318, 1, 1993.

98. **Palmer, L. G., Li, J. H. Y., Lindeman, B., Edelman, I. S.,** Aldosterone control of the density of sodium channels in the toad urinary bladder, *J. Membr. Biol.,* 64, 91, 1982.

99. Cuthbert, A. W., Shum, W. K., Effects of vasopressin and aldosterone on amiloride binding in toad epithelial cells, *Proc. R. Soc. Lond. (Biol.),* 189, 543, 1975.

100. **Szerlip, H., Pavlesky, P., Cox, M., Blazer-Yost, B.,** Relationship of the aldosterone-induced protein, GP70, to the conductive Na^+ channel, *J. Am. Soc. Nephrol.,* 2, 1108, 1991.

101. **Garty, H., Edelman, I. S.,** Amiloride-sensitive trypsinization of apical sodium channels. Analysis of hormonal regulation of sodium transport in toad bladder, *J. Gen. Physiol.,* 81, 785, 1983.

102. **Kleyman, T. R., Cragoe, E. J., Jr., Kraehenbühl, J. P.,** The cellular pool of Na^+ channels in the amphibian cell line A6 is not altered by mineralocorticoids. Analysis using a new photoactive amiloride analog in combination with anti-amiloride antibodies, *J. Biol. Chem.,* 264, 11995, 1989.

103. **Kleyman, T. R., Coupaye-Gerard, B., Ernst, S. A.,** Aldosterone does not alter apical cell-surface expression of epithelial Na^+ channels in the amphibian cell line A6, *J. Biol. Chem.,* 267, 9622, 1992.

104. **Palmer, L. G., Corthésy-Theulaz, I., Gaeggeler, H. P., Kraehenbühl, J. P., Rossier, B. C.,** Expression of epithelial Na channels in Xenopus oocytes, *J. Gen. Physiol.,* 96, 23, 1990.

105. **Palmer, L. G., Pacha, J., Frindt, G.,** Regulation of apical sodium channels in the rat cortical collecting tubule by aldosterone, in *Aldosterone: Fundamental aspects,* Bonvalet, J. P., Farman, N., Lombès, M., Rafestin-Oblin, M. E., Eds., Colloque INSERM/John Libbey Eurotext Ltd., Paris, 1991, 285.

106. **Sweadner, K. J.,** Isozymes of Na^+/K^+-ATPase, *Biochim. Biophys. Acta,* 988, 185, 1989.

107. **Doucet, A.,** Function and control of Na-K-ATPase in single segments of the mammalian nephron, *Kidney Int.,* 34, 749, 1988.

108. **Doucet, A., Katz, A. I.,** Short-term effects of aldosterone on Na-K-ATPase in single nephron segments, *Am. J. Physiol.,* 241, F273, 1981.

109. **El Mernissi, G., Doucet, A.,** Short-term effects of aldosterone and dexamethasone on Na-K-ATPase along the rabbit nephron, *Pflügers Arch.,* 399, 147, 1983.

110. **Barlet-Bas, C., Khadouri, C., Marsy, S., Doucet, A.,** Enhanced intracellular sodium concentration in kidney cells recruits a latent pool of Na-K-ATPase whose size is modulated by corticosteroids, *J. Biol. Chem.,* 265, 7799, 1990.

111. **Blot-Chabaud, M., Wanstok, F., Bonvalet, J. P., Farman, N.,** Cell sodium-induced recruitment of Na^+-K^+-ATPase pumps in rabbit cortical collecting tubules is aldosterone dependent, *J. Biol. Chem.,* 265, 11676, 1990.

112. **Jorgensen, P. L.,** Regulation of the $(Na^+ + K^+)$-activated ATP hydrolyzing enzyme system in rat kidney. I. The effect of adrenalectomy and the supply of sodium on the enzyme system, *Biochim. Biophys. Acta,* 151, 212, 1968.

113. **Rafestin-Oblin, M. E., Couette, B., Barlet-Bas, C., Cheval, L., Viger, A., Doucet, A.,** Renal action of progesterone and 18-substituted derivatives. *Am. J. Physiol.,* 260, F828, 1991.

114. **El Mernissi, G., Chabardès, D., Doucet, A., Hus-Citharel, A., Imbert-Teboul, M., Le Bouffant, F., Montégut, M., Siaume, S., Morel, F.,** Changes in tubular basolateral markers after chronic DOCA treatment, *Am. J. Physiol.,* 245, F100, 1983.

115. **Hayhurst, R. A., O'Neil, R.,** Time-dependent actions of aldosterone and amiloride on Na^+-K^+-ATPase of cortical collecting duct. *Am. J. Physiol.,* 254, F689, 1988.

116. **Barlet-Bas, C., Khadouri, C., Marsy, S., Doucet, A.,** Sodium-independent *in vitro* induction of Na^+, K^+-ATPase by aldosterone in renal target cells: Permissive effect of triiodothyronine, *Proc. Natl. Acad. Sci. U.S.A.,* 85, 1701, 1988.

117. **Barlet, C., Doucet, A.,** Triiodothyronine enhances renal response to aldosterone in the rabbit collecting tubule, *J. Clin. Invest.,* 79, 629, 1987.

118. **Geering, K., Girardet, M., Bron, C., Kraehenbühl, J. P., Rossier, B. C.,** Hormonal regulation of (Na^+, K^+)-ATPase biosynthesis in the toad bladder. Effect of aldosterone and 3,5,3'-triiodo-l-thyronine, *J. Biol. Chem.,* 257, 10338, 1982.

119. **El Mernissi, G., Doucet, A.,** Specific activity of Na-K-ATPase after adrenalectomy and hormone replacement along the rabbit nephron, *Pflügers Arch.,* 402, 258, 1984.

120. **Farman, N., Coutry, N., Logvinenko, N., Blot-Chabaud, M., Bourbouze, R., Bonvalet, J. P.,** Adrenalectomy reduces α_1 and not β_1 Na^+-K^+-ATPase mRNA expression in rat distal nephron, *Am. J. Physiol.,* 263, C810, 1992.

121. **Verrey, F., Kraehenbühl, J. P., Rossier, B. C.,** Aldosterone induces a rapid increase in the rate of Na,K-ATPase gene transcription in cultured kidney cells, *Mol. Endocrinol.,* 3, 1369, 1989.

122. **Verrey, F., Schaerer, E., Zoerkler, P., Paccolat, M. P., Geering, K., Kraehenbühl, J. P., Rossier, B. C.,** Regulation by aldosterone of Na^+, K^+, ATPase mRNAs, protein synthesis, and sodium transport in cultured kidney cells, *J. Cell Biol.,* 104, 1231, 1987.

123. **Johnson, J. P., Jones, D., Wiesmann, W. P.,** Hormonal regulation of Na^+-K^+-ATPase in cultured epithelial cells, *Am. J. Physiol.,* 251, C186, 1986.

124. **Leal, T., Crabbé, J.,** Effects of aldosterone on $(Na^+ + K^+)$-ATPase of amphibian sodium-transporting epithelial cells (A6) in culture, *J. Steroid Biochem.,* 34, 581, 1989.

125. **Ernst, S. A., Gaeggeler, H. P., Geering, K., Kraehenbühl, J. P., Rossier, B. C.,** Differential effect of aldosterone on expression of apical membrane proteins and of basolateral membrane β-subunit of Na, K-ATPase, *J. Cell Biol.,* 103, 465a, 1986.

126. **Rossier, B. C., Verrey, F., Kraehenbühl, J. P.,** Transepithelial sodium transport and its control by aldosterone: A molecular approach, *Curr. Top. Membr. Transp.,* 34, 167, 1989.

127. **Petty, K. J., Kokko, J. P., Marver, D.,** Secondary effect of aldosterone on Na-K-ATPase activity in the rabbit cortical collecting tubule, *J. Clin. Invest.,* 68, 1514, 1981.

128. **Oberleithner, H., Weigt, M., Westphale, H.-J., Wang, W.,** Aldosterone activates Na^+/H^+ exchange and raises cytoplasmic pH in target cells of the amphibian kidney, *Proc. Natl. Acad. Sci. U.S.A.,* 84, 1464, 1987.

129. **Barlet-Bas, C., Cheval, L., Féraille, E., Marsy, S., Doucet, A.,** Regulation of tubular Na-K-ATPase, in *Nephrology,* Hatano, M., Ed., Springer-Verlag, Tokyo, 1991, 419.

130. **Fimognari, G. M., Porter, G. A., Edelman, I. S.,** The role of the TCA cycle in the action of aldosterone on Na^+ transport. *Biochim. Biophys. Acta,* 135, 89, 1967.

131. **Feldman, D., Vander Wender, C., Kessler, E.,** The effect of aldosterone on oxidative enzymes of the rat kidney. *Biochim. Biophys. Acta,* 51, 401, 1961.

132. **Kirsten, E., Kirsten, R., Leaf, A., Sharp, G. W. G.,** Increased activity of enzymes of the tricarboxylic acid cycle in response to aldosterone in the toad bladder, *Pflügers Arch.,* 300, 213, 1968.

133. **Kirsten, R., Brinkhoff, B., Kirsten, E.,** Increase of cytochrome α and α_3 in rat kidney mitochondria in response to administration of aldosterone *in vivo. Pflügers Arch.,* 314, 231, 1970.

134. **Kirsten, R., Kirsten, E.,** Redox state of pyridine nucleotides in renal response to aldosterone, *Am. J. Physiol.,* 223, 229, 1972.

135. **Trachewsky, D.,** Aldosterone stimulation of riboflavin incorporation into rat renal flavin coenzymes and the effect of inhibition by riboflavin analogues on sodium reabsorption. *J. Clin. Invest.,* 62, 1325, 1978.

136. **Law, P. Y., Edelman, I. S.,** Induction of citrate synthase by aldosterone in the rat kidney, *J. Membr. Biol.,* 41, 41, 1978.

137. **Marver, D., Schwartz, M. J.,** Identification of mineralocorticoid target sites in the isolated rabbit cortical nephron. *Proc. Natl. Acad. Sci. U.S.A.,* 77, 3672, 1980.

138. **Johnson, J. P., Green, S. W.,** Aldosterone stimulates Na^+ transport without affecting citrate synthase activity in cultured cells. *Biochim. Biophys. Acta,* 647, 293, 1981.

139. **Sansom, S. C., O'Neil, R. G.,** Mineralocorticoid regulation of apical membrane Na and K transport of the cortical collecting duct, *Am. J. Physiol.,* 248, F858, 1985.
140. **Sansom, S., Muto, S., Giebisch, G.,** Na-dependent effects of DOCA on cellular transport properties of CCDs from ADX rabbits, *Am. J. Physiol.,* 253, F753, 1987.
141. **Schafer J. A., Troutman, S. H., Schlatter, E.,** Vasopressin and mineralocorticoid increase apical membrane driving force for K^+ secretion in rat CCD, *Am. J. Physiol.,* 258, F199, 1990.
142. **Schafer, J. A., Troutman, S. L.,** Potassium transport in cortical collecting tubules from mineralocorticoid-treated rat, *Am. J. Physiol.,* 253, F76, 1987.
143. **Frindt, G., Palmer, L. G.,** Low-conductance K channels in apical membrane of rat cortical collecting tubule. *Am. J. Physiol.,* 256, F143, 1989.
144. **O'Neil, R. G.,** Aldosterone regulation of sodium and potassium transport in the cortical collecting duct, *Semin. Nephrol.,* 10, 365, 1990.
145. **Sansom, S. C., O'Neil, R. G.,** Effects of mineralocorticoids on transport properties of cortical collecting duct basolateral membrane, *Am. J. Physiol.,* 251, F743, 1986.
146. **Stokes, J. B.,** Sodium and potassium transport by the collecting duct, *Kidney Int.,* 38, 679, 1990.
147. **Oberleithner, H.,** Aldosterone-regulated ion transporters in the kidney, *Klin. Wochenschr.,* 68, 1087, 1990.
148. **Wang, W., Henderson, R. M., Geibel, J., White, S., Giebisch, G.,** Mechanism of aldosterone-induced increase of K^+ conductance in early distal renal tubule cells of the frog, *J. Membr. Biol.,* 111, 277, 1989.
149. **Khadouri, C., Marsy, S., Barlet-Bas, C., Doucet, A.,** Effect of adrenalectomy on NEM-sensitive ATPase along rat nephron and on urinary acidification, *Am. J. Physiol.,* 253, F495, 1987.
150. **Garg, L. C., Narang, N.,** Effects of aldosterone on NEM-sensitive ATPase in rabbit nephron segments, *Kidney Int.,* 34, 13, 1988.
151. **Mujais, S. K.,** Effects of aldosterone on rat collecting tubule N-ethylmaleimide-sensitive adenosine triphosphatase, *J. Lab. Clin. Med.,* 109, 34, 1987.
152. **Khadouri, C., Marsy, S., Barlet-Bas, C., Doucet, A.,** Short-term effect of aldosterone on NEM-sensitive ATPase in rat collecting tubule, *Am. J. Physiol.,* 257, F177, 1989.
153. **Schwartz, G. J., Al Awqati, Q.,** Carbon dioxide causes exocytosis of vesicles containing H^+ pumps in isolated perfused proximal and collecting tubules, *J. Clin. Invest.,* 25, 1638, 1985.
154. **Garg, L. C.,** Effects of aldosterone on H-K-ATPase activity in rabbit nephron segments, in *Aldosterone: Fundamental aspects,* Bonvalet, J. P., Farman, N., Lombès, M., Rafestin-Oblin, M. E., Eds., Colloque INSERM/John Libbey Eurotext Ltd., Paris, 1991, 328.
155. **Khadouri, C., Barlet-Bas, C., Marsy, S., Cheval, L., Doucet, A.,** Heterogeneity of distribution and of regulation by corticosteroids of transport ATPases along the rat collecting tubule, in *Proceedings of the Conference on The Sodium Pump,* Bamberg, E., Schoner, W., Eds, Steinkopff Verlag, Darmstadt, in press.
156. **Garg, L. C., Narang, N.,** Supression of ouabain-insensitive K-ATPase activity in rabbit nephron segments during chronic hyperkaliemia, *Renal Physiol. Biochem.,* 12, 295, 1989.
157. **Cheval, L., Barlet-Bas, C., Khadouri, C., Féraille, E., Marsy, S., Doucet, A.,** K^+-ATPase-mediated Rb^+ transport in rat collecting tubule: modulation during K^+ deprivation, *Am. J. Physiol.,* 260, F800, 1991.
158. **Gifford, J. D., Rome, L., Galla, J. H.,** H^+-K^+-ATPase activity in rat collecting duct segments, *Am. J. Physiol.,* 262, F692, 1992.
159. **Oberleithner, H., Steigner, W., Silbernagl, S., Vogel, U., Gstraunthaler, G., Pfaller, W.,** Mardin-Darby canine kidney cells. III. Aldosterone stimulates an apical H^+/K^+ pump, *Pflügers Arch.,* 416, 540, 1990.
160. **Rajerison, R., Marchetti, J., Roy, C., Bockaert, J., Jard, S.,** The vasopressin-sensitive adenylate cyclase in the rat kidney: Effect of adrenalectomy and corticosteroids on hormonal receptor-enzyme coupling, *J. Biol. Chem.,* 249, 6390, 1974.

161. **Doucet, A., Barlet-Bas, C., Siaume-Perez, S., Khadouri, C., Marsy, M.,** Gluco- and mineralocorticoids control adenylate cyclase in specific nephron segments, *Am. J. Physiol.,* 258, F812, 1990.
162. **El Mernissi, G., Barlet-Bas, C., Khadouri, C., Cheval, L., Marsy, S., Doucet, A.,** Short-term effects of aldosterone on vasopressin-sensitive adenylate cyclase in rat collecting tubule, *Am. J. Physiol.,* 264, F821, 1993.
163. **Jackson, B. A., Braum-Werness, J. L., Kusano, E., Dousa, T. P.,** Concentrating defect in the adrenalectomized rat. Abnormal vasopressin-sensitive cyclic adenosine monophosphate metabolism in the papillary collecting duct, *J. Clin. Invest.,* 72, 997, 1983.

Chapter 3

THE EARLY RESPONSE TO ALDOSTERONE IN THE KIDNEY

Albrecht Schwab and Hans Oberleithner

TABLE OF CONTENTS

I. INTRODUCTION

Aldosterone is involved in the regulation of epithelial Na^+-, K^+- and H^+-transport in organs such as kidney, amphibian urinary bladder, frog skin, distal intestine, mammary gland, trachea, salivary gland, sweat gland and lung.[1–5] This review will be limited to studies performed in the kidney or in model epithelia such as the toad urinary bladder, frog skin or cultured cells. In particular the toad urinary bladder proved to be an extremely useful model and much of the present knowledge about the action of aldosterone was gained from this preparation.[6,7]

Experiments revealed a characteristic time course of the action of aldosterone. The acute application of aldosterone is followed by a lag time of 5 to 90 min in which transport properties of the epithelium are not yet affected. Following this lag period, Na^+ transport, measured, for example, as short circuit current (I_{sc}) across the toad bladder epithelium, often doubles and transepithelial resistance decreases. Until 3 to 4 h after application of aldosterone, this effect is referred to as an "early response". Thereafter, until 24 h after application of aldosterone, I_{sc} may increase further and this effect is called the "late response". "Chronic effects" of aldosterone are exerted when the hormone is applied for several days. During each period, specific cellular events are initiated and eventually lead to the physiological action of aldosterone. In this article we will exclusively focus on the "early response" of aldosterone, i.e., effects within the first 3 to 4 h after the application of the hormone.

Three main effects of aldosterone will be considered in detail in this article:

1. Na^+ transport
2. K^+ transport
3. H^+ transport

Table 1 provides an overview of the different techniques applied for studying the early response to aldosterone. At the end of each chapter we will briefly discuss which of the early effects of aldosterone comprises a *genomic* or a *nongenomic* action of the hormone.

II. EARLY EFFECTS OF ALDOSTERONE ON NA⁺ TRANSPORT

As identified by Koefoed-Johnson and Ussing in frog skin,[8] transepithelial Na^+ transport is a two-step process. Na^+ enters the cells passively along its electrochemical gradient via Na^+ selective ion channels in the apical cell membrane and is actively extruded out of the cell by the Na^+-K^+-ATPase located in the basolateral membrane. In the kidney, this two-step process results in Na^+ reabsorption. It is evident that there are two sites at which Na^+ reabsorption can be coordinately regulated; the apical entry step via Na^+ channels, and the basolateral exit step mediated by the Na^+-K^+-ATPase. Of course, Na^+ reabsorption has to be tightly regulated in the kidney. In particular, in distal portions of the nephron the fine tuning of the final urine composition is accomplished and aldosterone plays an important role in its regulation. Accordingly, the highest number of nuclear mineralocorticoid binding sites or mineralocorticoid receptors in the rabbit kidney were found in distal parts of the nephron.[9,10]

One of the major effects of aldosterone is to enhance tubular Na^+ reabsorption. This antinatriuretic action has been shown in many studies (see Reference 11 for further references). In the following, we will discuss the acute effect of aldosterone on Na^+ reabsorption. An early effect of aldosterone on Na^+ reabsorption was seen in adrenalectomized dogs where injection of aldosterone into the aorta or into the renal artery decreased Na^+ excretion within 5 to 30 min.[12] Similarly, adrenalectomized rats responded to injections or infusions of aldosterone with antinatriuresis, the onset of which was after a lag time of 30

TABLE 1
Techniques for Studying the Early Response to Aldosterone

Technique	Experimental model	Lag time	Ref.
Clearance	Dog, rat	5′–60′	12–14, 110, 124
Microperfusion	Rat	3 h	102, 110, 116
Short circuit current	Toad bladder, TBM cells Frog skin, turtle bladder	1–2 h	24, 30, 43, 97, 114
Microelectrode (conventional and ion-selective)	Rabbit CCD, toad bladder, rabbit urinary bladder frog diluting segment, MDCK cells, frog skin	10′–3 h	16, 17, 42–44, 57, 97, 98, 101, 115, 117, 119
Tracer flux	Rabbit CCD, toad bladder	≤3 h	18, 26, 30, 102
Noise analysis	Toad bladder	4 h	23
Vesicle studies	Toad bladder, A6 cells	3–5 h	30, 49
Patch clamp	Rat CCD, A6 cells	2–3 h	40, 41, 50
Electron microprobe	Toad bladder	4 h	58
RNA expression	Toad bladder	3 h	25, 26
Electrophoresis	Cultured CCD cells	1 h	125
Ouabain binding	A6 cells, rat CCD	3 h	70, 73–75
ATPase microassays	Rabbit, rat	3 h	59-67, 120
^{86}Rb uptake	Rat	30′	75
Membrane capacitance measurement	Frog skin	12′	122
Fluorometry	MDCK cells	10–20′	45
Atomic force microscopy	MDCK cells	6 h	131

to 60 min and reached a maximum after 2 to 3 h.[13,14] After 90 min, aldosterone had also stimulated protein and RNA synthesis. The rapid effect of aldosterone on Na^+ excretion depended on RNA synthesis since actinomycin D blunted antinatriuresis.[13] Similar effects of inhibitors of RNA and protein synthesis were also found in the toad bladder. The early aldosterone-induced increase of I_{SC} was prevented by these drugs.[15] Moreover, in isolated perfused cortical collecting ducts of rabbits whose endogenous aldosterone secretion was suppressed by high-salt diet, acute application of the hormone led to the rapid development of a lumen-negative transepithelial potential. After a latency of 10 to 20 min, the transepithelial potential rose from approximately 0 mV to -19 mV. In the presence of the aldosterone-antagonist spironolactone, aldosterone failed to change the transepithelial potential.[16] In a preceding study, the same group had demonstrated that the transepithelial potential depended on active Na^+ transport.[17] Hence, the aldosterone-induced rise of transepithelial potential reflected increased Na^+ transport in the cortical collecting duct.

However, these studies could not solve the question whether aldosterone induced antinatriuresis by stimulating Na^+ channels in the apical membrane (Na^+ entry step), or by activating the Na^+-K^+-ATPase in the basolateral membrane (Na^+ exit step).

We will now discuss experiments that were designed to elucidate the role aldosterone plays either in regulating apical Na^+ channels or the basolateral Na^+-K^+-ATPase.

A. EARLY EFFECTS OF ALDOSTERONE ON NA[+] CHANNELS

Using the toad bladder preparation, Crabbé[18] could demonstrate that aldosterone increased cellular $^{22}Na^+$ uptake at the same time short circuit current was stimulated by the hormone. Crabbé concluded that aldosterone stimulated Na^+ channels in the apical plasma membrane and thereby increased the availability of Na^+ for the basolateral Na^+-K^+-ATPase. This concept was confirmed in experiments in which elevated intracellular Na^+ availability, brought about by permeabilizing the apical plasma membrane with the cationic ionophore amphotericine B, also stimulated Na^+ transport.[19,20]

However, it was more than a decade until the effect of aldosterone on the apical cell membrane of frog skin (another model for Na^+ transporting epithelia), could be directly shown by the use of microelectrodes.[21] The major effect of 12 h pretreatment with aldosterone was to increase apical membrane Na^+ conductance by more than threefold. In addition, the conductance of the basolateral membrane was increased by 80%, i.e., aldosterone stimulated Na^+ transport by a coordinated dual action on apical and basolateral membranes.

Using noise analysis techniques, it was possible to identify Na^+ channels in frog skin[22] and to correlate the effect of aldosterone on toad bladder I_{SC} with the activity of Na^+ channels in the apical cell membrane.[23] As shown many times before, aldosterone doubled I_{SC} within 4 to 6 hours. This rise in macroscopic electrical current was the result of an increase in the number of conducting Na^+ channels in the apical membrane. The current through each individual Na^+ channel was calculated not to be altered. The authors proposed that following protein synthesis, there was a reversible recruitment of preexisting Na^+ channels from a reservoir of electrically undetectable Na^+ channels.

Trypsinization of the apical membrane of toad bladders in the presence or in the absence of the specific Na^+ channel blocker amiloride also revealed that aldosterone activated pre-existing Na^+ channels.[24] Trypsinization of the apical membrane of toad bladders prior to the application of aldosterone, greatly diminished the stimulation of I_{SC}. This loss of aldosterone response, however, could be rescued, when amiloride was present during trypsinization. Hence, Na^+ channels that were recruited by aldosterone[23] were already present in the apical cell membrane before the hormone was added. Moreover, these experiments ruled out that the early effect of aldosterone was due to *de novo* synthesis of Na^+ channels.

The studies by Palmer and Garty addressed one of the principal questions concerning the mechanism by which aldosterone increased apical Na^+ permeability. Does aldosterone change the total pool of Na^+ channels by *de novo* synthesis, or does aldosterone only change the pool of conducting Na^+ channels? Several groups investigated this question in recent years by expressing

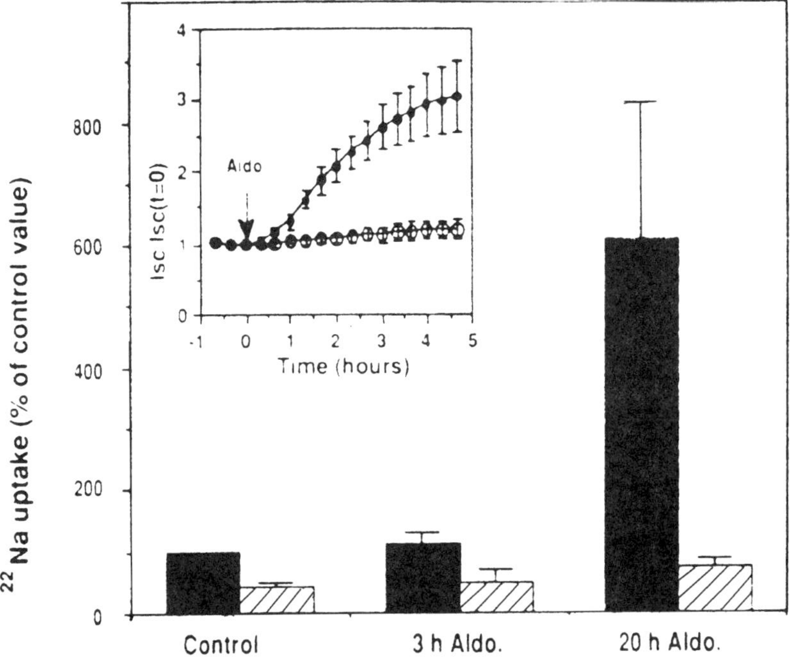

Figure 1: Expression of Na⁺ channels from turtle bladder after incubation with aldosterone. Total RNA was injected into oocytes that were assayed 3 d later for ^{22}Na⁺ uptake, in the presence *(hatched bars)*, and in the absence *(filled bars)*, of amiloride. *Inset:* I_{SC} of paired hemibladders monitored under control conditions and after stimulation with aldosterone (0.5 µmol/l). (From Asher, C., Eren, R., Kahn, L., Yeger, O., Garty, H., *J. Biol. Chem.*, 267, 16061, 1992. With permission.)

amiloride-blockable Na⁺ channels from control vs. aldosterone-stimulated cells,[25,26] or by using specific labeling techniques for this Na⁺ channel.[27–29]

Expression of total RNA from toad bladders in Xenopus oocytes gave rise to amiloride-sensitive ^{22}Na⁺ uptake.[26] As would be expected from the above mentioned studies, preincubating bladders with aldosterone 3 h prior to the isolation of RNA did not increase ^{22}Na⁺ uptake into oocytes although I_{SC} of the bladder had doubled by that time. Amiloride-sensitive ^{22}Na⁺ uptake only rose after a 20 h pre-incubation period with aldosterone (Figure 1). The authors concluded that two mechanisms had to come into play in order to explain the different effects of short term and long term exposure to aldosterone on the level of transcription. Apparently, only the late effect of aldosterone could be accounted for by induction of new channels. These findings would also explain earlier results from this group.[30] The stimulatory effect of aldosterone on Na⁺ transport was not retained in isolated vesicles when they were from tissue which had been incubated with the hormone for 3 h. This was only the case after a 6 h incubation period.

Expression of Na^+ channels from A6 cells, a renal cell line derived from Xenopus kidney,[31] depended on culture conditions.[25] Expression of Na^+ channels in oocytes after injection of mRNA from A6 cells was not affected even by preincubating A6 cells for 24 h with aldosterone when they were grown on permeable supports, i.e., under conditions of active transport. According to our definition, preincubating cells for 24 h with aldosterone will reflect a "late response" to the hormone. But even after this time, there were no signs for increased transcriptional activity for Na^+ channels of filter grown cells. However, when cells were grown on an impermeable support, i.e., under conditions in which transepithelial Na^+ transport was confined to dormes and cells presumably were not as differentiated as on permeable supports, aldosterone increased transcription of Na^+ channels in A6 cells. After this pretreatment, Na^+ channel expression, after injection of mRNA into oocytes, was enhanced, but the level of expression was much lower than in cells grown on permeable supports.[25] Nevertheless, it is safe to conclude that Na^+ channels activated during the "early response" to the hormone are not aldosterone-induced proteins.

Experiments using specific labeling techniques with antibodies or amiloride analogues in aldosterone-stimulated A6 cells came to the same results.[27–29] The pool of Na^+ channels was neither altered by the hormone, nor by its antagonist spironolacton.

If *de novo* protein synthesis of Na^+ channels was excluded, then what is the cause for the early increase of Na^+ permeability of the apical membrane after the addition of aldosterone? The introduction of patch clamp techniques[32] gave more insight to this important question. With these techniques, the properties and the regulation of Na^+ channels from rat cortical collecting duct principal cells, or from cultured A6 cells could be characterized in great detail (for reviews see references 33–38). Their most important properties are low single channel conductance, high selectivity for Na^+, and sensitivity to amiloride.

Taking the single channel properties into account, the Na^+ conductance G_{Na} of the apical cell membrane is given by the following equation:

$$G_{Na} = g_{Na} \cdot N \cdot p_o \tag{1}$$

g_{Na} is the single channel conductance, N the number of channels present in the apical cell membrane, and p_o is the open probability, a measure for the time the channel spends in the open state.[39] Since the single channel conductance is not affected by aldosterone,[23,40] g_{Na} can increase when either the number of channels (N) rises (channel recruitment), or when the open probability (P_o) rises (activation of channels). Depending on the tissue studied, there is experimental evidence for both possibilities.

A6 cells were studied in the presence of aldosterone, in the aldosterone-depleted state, and after the acute re-addition of the hormone.[40] Upon aldosterone removal, the product of channel number per patch and open probability (Np_o) had decreased after 36 h. However, upon re-addition of aldosterone to hormone-depleted cells, Np_o rapidly rose within 2 h. In order to assess whether the rise of Np_o was related to altered channel number or open probability, the authors analyzed the maximal number of channels present in the patch. Based on this statistical analysis, they concluded that aldosterone primarily changed open probability p_o, i.e., Na^+ channels were activated rather than recruited from a silent pool.

In contrast, studies performed on Na^+ channels from rat cortical collecting duct (CCD) principal cells came to a different conclusion.[41] Since, in normally fed animals, virtually no Na^+ channel activity was present, the appearance of Na^+ channels was either induced by feeding the animals with a Na^+-depleted diet or by injecting furosemide into control animals. Whereas the dietary regimen depended on a chronic effect of aldosterone — Np_o closely correlated with plasma aldosterone levels — the injection of furosemide was aimed to elicit an early response. The drug resulted in rapid volume depletion and aldosterone secretion. 3 h after injection of furosemide, channel activity could be observed. Based on histograms depicting the number of channels per patch, it was concluded that aldosterone increased channel number and not open probability, i.e., Na^+ channels were recruited rather than being activated.[41]

From these studies the next question immediately arises. What is the molecular mechanism responsible for activating or recruiting Na^+ channels after the addition of aldosterone? Several possibilities of posttranslational modification have been suggested.

Aldosterone is known to increase intracellular pH within 20 to 90 min in renal cells from the amphibian kidney,[42] in frog skin,[43] and in MDCK cells.[44,45] Moreover, in frog skin epithelial cells a rise of Na^+ transport was accompanied by a concomitant intracellular alkalinization suggesting a physiological role of pH_i in the control of transepithelial Na^+ transport.[43] In CCD principal cells of chronically Na^+-depleted rats, such an effect of aldosterone could not be observed. However, this does not exclude the possibility that hormone-induced alkalinization occurred at an earlier time and dissipated in the chronic state.[41] At least in the inside-out patch configuration Na^+ channels from rat CCD were quite sensitive to alterations of pH on their cytoplasmic side.[46] Acidosis reduced channel activity.

Another possible regulatory step could be methylation of phospholipids or of proteins. In cultured toad bladder epithelial cells, aldosterone stimulated protein methylation and transmethylation reactions. Inhibition of aldosterone-induced methylations blocked the early response to aldosterone on Na^+ transport.[47] In cultured renal collecting duct cells, protein methylation was observed as early as 1 h after administration of aldosterone. At this time initial aldosterone-dependent electrical changes also occurred.[48] Similarly, a stimulatory effect of protein and phospholipid methylation on Na^+ transport was found in isolated apical membrane vesicles from A6 cells. Accordingly, in vesicles from aldosterone-treated cells (5 h), methylation did not stimulate Na^+ transport any further.[49] In preliminary experiments, the stimulatory effect of methylation on Na^+ channel activity could be reproduced on the single channel level.[50] However, when expressing the Na^+ channel from A6 cells in oocytes, the sensitivity to methylation reaction seems to be lost.[26] Therefore, it still remains to be clarified which proteins are methylated: is it the Na^+ channel itself; what role does methylation play in mediating the early aldosterone response; and of course by which mechanism does aldosterone cause methylation at all?

We will now consider one final point in the regulation of the apical membrane Na^+ channel by aldosterone: the synergism between aldosterone and vasopressin (AVP). Like aldosterone, AVP also stimulated Na^+ transport in rat and rabbit CCD[51] and in toad bladder. Simultaneous addition of the two hormones to toad bladders potentiated their action on Na^+ transport within 60 min. This effect could be mimicked by simultaneous addition of aldosterone and cAMP.[52] AVP alone activates the Na^+ channel via cAMP-dependent phosphorylation of specific channel subunits[53] (for review see References 33, 35–37, 54).

In order to explain the synergism between aldosterone and AVP, the following model was proposed[36] (Figure 2). Na^+ channels are present in two pools, one in the membrane and one "inaccessible" possibly located in subapical vesicles. Aldosterone is converting Na^+ channels in both pools from an inactive to an active form. AVP, on the other hand, promotes insertion of Na^+ channels from the "inaccessible" pool into the plasma membrane. Accordingly, after pretreatment with aldosterone the effect of AVP on Na^+ transport is enhanced, since Na^+ channels will now be inserted in their active form.* Additionally, the early stimulatory effect of aldosterone on AVP-sensitive adenylate cyclase[55] may also come into play for the synergistic action of aldosterone and AVP.

In summary, aldosterone causes an early increase of apical Na^+ conductance (Na^+ entry step) by activating preexisting Na^+ channels. However, *de novo* synthesis of the Na^+ channel protein itself was excluded. Although protein methylations or intracellular pH shifts have been shown to be involved in channel activation, the exact molecular mechanism by which aldosterone causes early channel activation still remains to be elucidated.

1. Genomic vs. Nongenomic Effects of Aldosterone

The question whether aldosterone stimulates apical Na^+ uptake via genomic or nongenomic effects is not yet conclusively solved. The lack of Na^+ channel induction during the early response phase[26,30] would be consistent with a nongenomic action of the hormone. However, these studies did not rule out the possibility that aldosterone induces the activation of a signaling network of mixed genomic and nongenomic effectors. Such a signaling chain might consist, for example, of genomic activation of the Na^+/H^+ exchange[42] followed by nongenomic stimulation of pH dependent processes such as the activation of Na^+ channels.

---

* According to recent experiments,[53] this model has to be modified in so far as it was suggested that AVP activates Na^+ channels already present in the membrane rather than causing exocytosis of subapical vesicles.

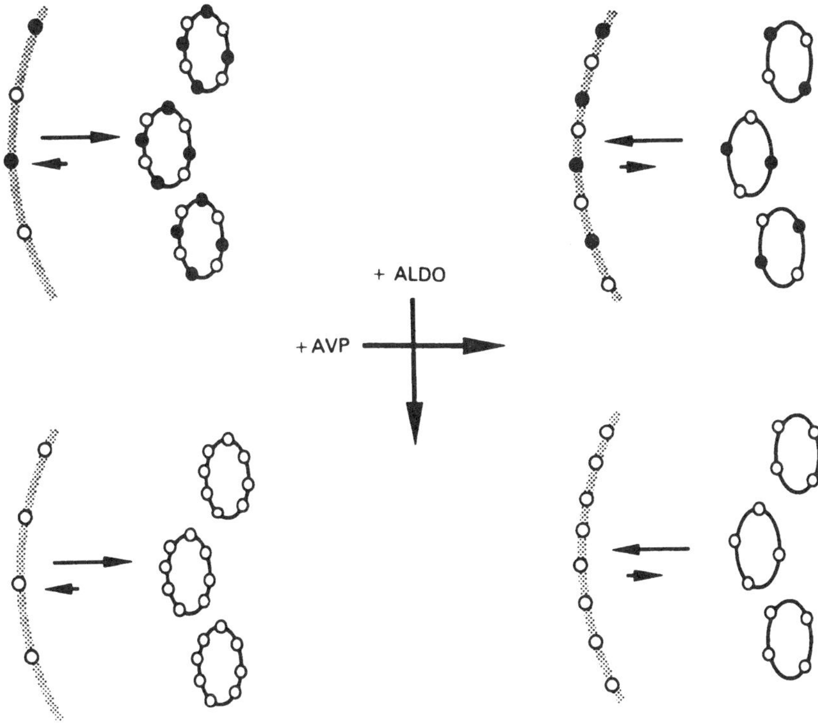

Figure 2: Model to explain the synergism between AVP and aldosterone in stimulating Na^+ reabsorption in Na^+ transporting epithelia. Inactive channels shown by closed circles, active channels shown by open circles. (From Schafer, J. A., Hawk, C. T., *Kidney Int.*, 41, 255, 1992. Reprinted by permission of Blackwell Scientific Publications, Inc.)

B. EARLY EFFECTS OF ALDOSTERONE ON NA⁺-K⁺-ATPase

In the preceding section we described how aldosterone activates Na^+ channels in the apical membrane. In order to avoid intracellular Na^+ accumulation, the extrusion of Na^+ by the Na^+-K^+-ATPase located in the basolateral membrane must also be activated. Theoretically, this could be achieved in two ways: the pump rate of the Na^+-K^+-ATPase is increased either due to the elevated intracellular Na^+ load, and/or the Na^+-K^+-ATPase itself is also a target site for the regulation by aldosterone.

Early evidence for a coordinate dual action of aldosterone came from experiments using the toad bladder preparation where aldosterone did not change the concentrations of intracellular Na^+, K^+, and Cl^- at a time of stimulated I_{SC}.[56] However, small changes (<10%) might have been overlooked in this study. In later studies, intracellular Na^+ was measured in single cells either by the use of Na^+-selective microelectrodes in rabbit urinary bladder cells,[57] or by quantitative electron microprobe analysis in toad urinary bladder cells.[58] In contrast to the aforementioned results, in both preparations aldosterone caused a rapid (1 to 4 h) increase of intracellular Na^+ from 7 to 20 mmol/l and from 9 to 13 mmol/l, respectively. Nonetheless, the authors of the latter study interpreted their results as not contradictory to the proposed dual action of aldosterone since the rise of intracellular Na^+ was much less than after addition of AVP.[58]

The early effect of aldosterone on basolateral Na^+-K^+-ATPase was discussed controversially. In particular, the duration of the latent period varied considerably. In adrenal-intact animals, mineralocorticoid treatment caused no early effects on Na^+-K^+-ATPase activity, whereas in adrenalectomized animals, aldosterone was seen to rapidly restore Na^+-K^+-ATPase activity.[59–64] Others, however, failed to demonstrate this early effect.[65–67] A possible explanation for the apparent variability of the latent period is given by Hayhurst and O'Neil.[68] They proposed that the initial level of Na^+-K^+-ATPase activity determines whether or not aldosterone will elicit an early response. Reviewing the literature, they found a linear relationship between the latency period for aldosterone action on the Na^+-K^+-ATPase and the initial Na^+-K^+-ATPase activity, i.e., low basal levels of Na^+-K^+-ATPase activity after adrenalectomy (20% of control), were a prerequisite for a rapid onset of the action of aldosterone. When initial Na^+-K^+-ATPase activity was higher (>50%) the onset would be delayed.

We next have to address the questions whether aldosterone increases the pump rate of already existing pumps, whether this process depends on an elevated intracellular Na^+ load, and/or whether aldosterone also increases the number of pumps. Several micromethods have been developed in order to determine Na^+-K^+-ATPase activity. In particular, one has to distinguish whether an increase of V_{max} of this enzyme is due to elevated enzyme activity, due to an elevated number of active pump units, or due to a combination of both. Available techniques to quantify enzyme activity include measurements of ^{32}P-ATP hydrolysis, fluorimetric methods which couple the rate of ATP hydrolysis to fluorescence of NAD/NADH or NADP/NADPH, and ouabain-sensitive initial ^{86}Rb uptake. With the latter technique, pump activity is not studied under V_{max} conditions, so that changes in pump activity due to transient increases of intracellular Na^+ can be also detected. The number of Na^+-K^+-ATPase units can be determined by specific ^3H-ouabain binding.[69]

A widely used approach to look for Na^+-induced changes of Na^+-K^+-ATPase activity is to block Na^+ entry through apical membrane Na^+ channels by amiloride. In adrenalectomized rats, injection of this drug prior to substitution with aldosterone prevented the restoration of Na^+-K^+-ATPase activity. Accordingly, the authors concluded that acute changes in Na^+-K^+-ATPase activity observed in the absence of amiloride were a secondary effect due to aldosterone-induced stimulation of apical Na^+ entry.[61] Alternatively, intracellular Na^+ concentration can be elevated by the use of cationic ionophores such as nystatin or amphotericine B. Using this technique, it was found in rat and rabbit CCDs that a rise in intracellular Na^+ results in the recruitment of a latent pool of Na^+-K^+-ATPase units. Both activity and specific ouabain-binding increased. *De novo* synthesis of the pump was ruled out, since actinomycin D or cycloheximide were without effect. In adrenalectomized animals, this pool of Na^+-K^+-ATPase units was no longer present, pointing to the long term control by aldosterone.[70,71] However, the time-course of Na^+-induced Na^+-K^+-ATPase recruitment was fast.[72] Pump recruitment was fully achieved within 1 to 2 min as revealed by a rapid increase of ^{3}Houabain binding to rabbit CCDs. Interestingly, this effect was not affected by colchicine or cytochalasin B, suggesting activation of pumps already present in the membrane. Restoration of a low intracellular Na^+ concentration was paralleled by a decrease of ^{3}H-ouabain binding, i.e., pump recruitment was a reversible phenomenon.[72]

Hence, Na^+-dependent recruitment of Na^+-K^+-ATPase units offers the ability to rapidly adjust pump rate to aldosterone-induced elevations of intracellular Na^+. However, as we discussed in the preceding section on the Na^+ entry step, aldosterone-induced stimulation of apical Na^+ channels requires at least 1 h to occur. Moreover, stimulation of the Na^+/H^+ exchanger by aldosterone mediating Na^+ influx into cells, too, was apparent in MDCK cells only after 10 to 20 min.[45] Therefore, it remains to be clarified whether such a rapid recruitment of Na^+-K^+-ATPase is crucial for the physiological response to aldosterone.

In addition to Na^+-dependent recruitment of Na^+-K^+-ATPase units, direct effects of aldosterone on pump number and activity have also been reported. In A6 cells aldosterone increased the number of Na^+-K^+-ATPase units within 3 h.[73] In CCDs from adrenalectomized rats the same effect was observed when T_3 was added.[74] Abolition of this early response by actinomycin D and cycloheximide pointed to protein synthesis.[73,74] Similar findings were made in adrenal-intact animals by other authors[75] who also investigated whether aldosterone elicited early alterations in the function of the pump by measuring ouabain-sensitive initial ^{86}Rb uptake. They found aldosterone to stimulate ^{86}Rb uptake into rat CCD cells within 30 min.[75] Aldosterone-induced changes of intracellular pH[42] were discussed as one possible mechanism for this effect since Na^+-K^+-ATPase activity is known to be very pH-sensitive.[76,77]

Early effects of aldosterone could also be seen on the transcriptional level.[78] Transcription rates of Na^+-K^+-ATPase subunit genes were found to be elevated as early as 15 min after the addition of aldosterone to A6 cells, i.e., at a time when Na^+ transport was not yet activated. Hence, transcriptional changes were clearly independent of intracellullar Na^+. Interestingly, aldosterone-induced stimulation of transcription could be blunted by cycloheximide. This finding suggested that aldosterone acted indirectly on Na^+-K^+-ATPase gene expression or that the hormone-receptor complex on the Na^+-K^+-ATPase gene promoter required the presence of a short-lived protein.[78] Although transcription of Na^+-K^+-ATPase subunits was very rapidly enhanced by aldosterone, targeting of newly synthesized Na^+-K^+-ATPase units to the basolateral membrane appeared to be a slow process and was accomplished only in the late phase of the response to the hormone.[79–81]

In MDCK cells the early response to aldosterone was further analyzed and divided into two successive periods. Initially, (<1 h), presynthesized Na^+-K^+-

ATPase units were inserted into the basolateral membrane independently from protein synthesis. After that time, increase in pump number relied on protein synthesis.[82] However, from the aforementioned study it seems unlikely that newly synthesized Na^+ pumps were inserted into the basolateral membrane within this short time. It is more likely that a regulatory protein was induced, which controlled recruitment and activation of pre-existing Na^+-K^+-ATPase units.

Taken together, these results show that aldosterone may regulate Na^+-K^+-ATPase activity in the basolateral membrane by a number of different mechanisms. Intracellular Na^+ is of great importance, but aldosterone may also stimulate pump activity in a Na^+-independent way. In contrast to apical Na^+ channels, the basolateral Na^+-K^+-ATPase is an aldosterone-induced protein whose transcription is initiated as early as 15 min after hormone addition. Nonetheless, the early response to aldosterone does not depend on newly synthesized Na^+-K^+-ATPase, but is mediated by increased activity and increased number of pumps.

One point that we have neglected so far is cell metabolism. Since Na^+ transport is an active process, its stimulation by aldosterone depends on appropriate energy supply.[83] Accordingly, in substrate-depleted toad bladders, aldosterone failed to stimulate Na^+ transport.[84] After stimulation with aldosterone, ATP consumption will rise due to increased Na^+-K^+-ATPase activity. One of the key enzymes involved in the generation of ATP via citric acid cycle and oxidative phosphorylation appears to be citrate synthase. In several studies, early aldosterone-dependent stimulation of citrate synthase and other enzymes from the tricarboxylic acid cycle were detected.[85–87] Accordingly, intracellular concentrations of ATP and ATP/ADP x P_i ratio correlated with I_{SC} during the early phase of the aldosterone response in toad bladders.[88] The increased level of energy-providing enzymes therefore appears to be complementary to increased Na^+ transport. That metabolism may control Na^+ transport was shown in toad bladders where inhibition of metabolism reduced Na^+ permeability.[89] Possibly, this was due to feedback-regulation of Na^+ channels with respect to pump activity, as demonstrated in rat CCD where inhibition of the pump reduced channel activity.[90]

The role of citrate synthase, and thereby of the metabolic coupling, was questioned by experiments performed on A6 cells in which aldosterone increased Na^+ transport without any detectable effect on citrate synthase activity.[91] But, it has to be kept in mind that in A6 cells, aldosterone-induced stimulation of Na^+ transport is mediated via receptors with low mineralocorticoid affinity[92] and not via high affinity receptors which are thought to be responsible for the early response to aldosterone.[81]

1. Genomic vs. Nongenomic Effects of Aldosterone

Aldosterone-dependent induction of the Na^+-K^+-ATPase is of a genomic nature. However, the time course of this process requires further discussion. *De novo* synthesis of the Na^+-K^+-ATPase was not accomplished during the early response phase.[79–81] Nonetheless, the rapid rise of pump number and activity involved genomic effects of aldosterone since it depended on protein synthesis.[73,74,82] In addition, transcription of the Na^+-K^+-ATPase itself also relied on genomic regulation.[78] Hence, we have to assume the activation of several genes controlling both synthesis and regulation of the Na^+-K^+-ATPase. As we pointed out, increased Na^+ transport was accompanied by a stimulation of energy-providing enzymes[86,87] which also reflected a genomic response to aldosterone. However, cell metabolism offers the possibility for secondary nongenomic regulation of Na^+ transport.[89,90]

Taken together, it appears that regulation of Na^+ transport by aldosterone involves a complicated network of genomic and nongenomic effects. Whereas genomic regulation of Na^+-K^+-ATPase has been directly demonstrated,[78] evidence for genomic regulation of other proteins is indirect and is based on the inhibitory effect of the aldosterone antagonist spironolacton, or on the effect of drugs such as actinomycin D or cycloheximide.

III. EARLY EFFECTS OF ALDOSTERONE ON K^+ TRANSPORT

In addition to its antinatriuretic action, aldosterone also promotes K^+ secretion[12–14,93] (see References 94, 95 for further references). There are several links by which the stimulatory action of aldosterone on Na^+ transport will indirectly enhance K^+ secretion. Enhanced Na^+ entry elicits a depolarization of the apical membrane,[57,96–98] thereby increasing the electrochemical driving force for cell-to-lumen K^+ secretion. Additionally, aldosterone-induced stimulation of the Na^+-K^+-ATPase will alter the ratio of intracellular ATP/ADP in favor of ADP. In patch clamp studies, such a shift was shown to activate the secretory apical K^+ channel in rat cortical collecting duct principal cells.[99,100] In this scenario, the ratio of ATP/ADP would mediate the "cross-talk" between apical and basolateral membranes and coordinate K^+ transport.

However, in addition to these indirect effects of aldosterone on K^+ secretion, there is good evidence that the hormone also regulates K^+ secretory pathways specifically and independently of Na^+ transport. Na^+ reabsorption and K^+ excretion could be dissociated in rats by the application of actinomycin D. Whereas aldosterone-induced kaliuresis was not blocked, in the same animals, hormone-dependent antinatriuresis was inhibited by this drug.[13] Microelectrode studies showed that aldosterone acutely increased the K^+ permeability of the luminal membrane from rat distal tubules,[101] a finding which was confirmed in rabbit CCD by the use of ^{42}K tracer measurements.[102] In the latter experiments, amiloride was added to the luminal fluid in order to eliminate Na^+ transport related processes.

Except for the implication that aldosterone-induced K^+ secretion was independent of protein synthesis, these studies left open the question as to the mechanism underlying the early stimulatory effect of aldosterone on kaliuresis. This question was addressed in studies performed on frog distal tubules ("diluting segment") and on MDCK cells.[42,44,103,104] These experiments were based on the observations that H^+ and K^+ secretion were functionally linked and could be induced by chronically exposing the animals to a high K^+ environment, a condition which was accompanied by elevated plasma aldosterone levels.[105–107] In fused tubule cells, aldosterone stimulated the apical Na^+/H^+ exchanger with a lag period of 20 min. After 1 h, intracellular pH was alkalinized by almost 0.3 pH units and pH recovery was accelerated following an acid load in an amiloride-sensitive way. In the intact tubule, aldosterone induced amiloride-sensitive acidification of the luminal fluid and approximately doubled tubular K^+ concentration.[42] Based on these results the following sequence of events was suggested: aldosterone activates the apical Na^+/H^+ exchanger and the ensuing intracellular alkalinization triggers K^+ channels in the apical membrane (see Figure 3). Moreover, in as yet unpublished experiments, aldosterone was shown to shift the setpoint of the Na^+-H^+-exchanger in frog diluting segment within 20 min from pH 6.99 to pH 7.33.[134]

This model was confirmed by several further studies. Using patch clamp techniques, the pH-sensitivity of apical K^+ channels in distal tubules could be directly demonstrated.[100,103,104] In MDCK cells and in frog skin principal cells, a similar amiloride-sensitive alkalinization of the cytoplasm was seen 10 to 20 min after the addition of aldosterone.[44,45,108] Most importantly, in the context of aldosterone-induced K^+ secretion, K^+ conductance of MDCK cells was very pH-sensitive in a range of intracellular pH which was controlled by aldosterone.[44] In frog skin, intracellular alkalinization activated a pH- and ATP-sensitive K^+ channel in the basolateral membrane whose role was to recycle K^+ taken up by the Na^+-K^+-ATPase.[108] Thus, intracellular pH could be a "second messenger" of aldosterone.

As mentioned above, aldosterone does not only affect the K^+ conductance of the apical cell membrane, but also that of the basolateral membrane. In TBM cells, a cell line derived from toad urinary bladder, the hormone also augmented the conductance of the basolateral membrane within 4 h.[97] This rise was independent of Na^+ transport and it could be ascribed to a rise of K^+ conductance. Such an increase is necessary for maintaining ionic homeostasis since, in situations of enhanced Na^+ transport, the active extrusion of Na^+ ions via the Na^+-K^+-ATPase is automatically followed by enhanced K^+ uptake. Hence, the physiological role for increased K^+ conductance was to allow sufficient K^+ "recycling" across the basolateral membrane. Therefore, parallel conclusions may be drawn to other epithelia, such as renal proximal tubule cells, in which enhanced Na^+ transport was also accompanied by increases of basolateral K^+ conductance. In renal proximal tubules, basolateral K^+ conductance and Na^+-K^+-ATPase were coupled by intracellular ATP.[109] Such metabolic coupling might also have come into play in frog skin principal cells where a basolateral ATP-sensitive K^+ channel was activated in less than 20 min after addition of aldosterone. The channel was regulated in part by the ratio of intracellular ATP/ADP, which decreased within 10 min after addition of the hormone.[108]

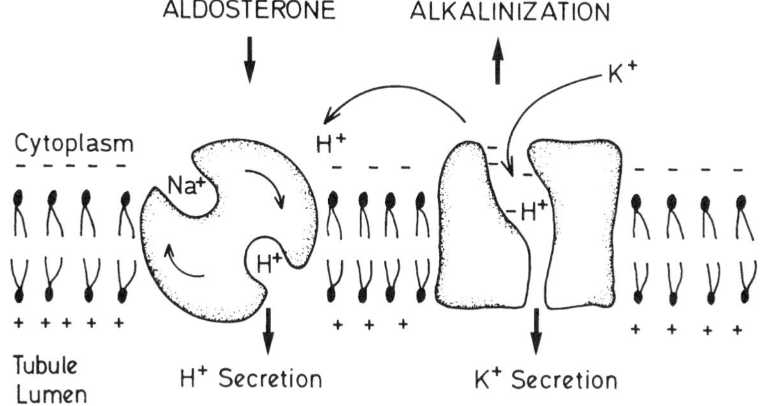

Figure 3:   Model of the action of aldosterone in cells of the amphibian diluting segment. pH sensitive K+ channels are functionally linked to the Na+/H+ exchanger. Intracellular alkalinization brought about by aldosterone-dependent activation of the exchanger enhances K+ secretion. (From Oberleithner, H., Weigt, N., Westphale, H.-J., Wang, W., *Proc. Natl. Acad. Sci. U.S.A.,* 84, 1464, 1987. With permission.)

However, *in vivo* renal K+ secretion is not only regulated on a single cell level, but factors such as tubular flow rate are also of paramount importance. It is well known that urinary K+ excretion directly correlates with urinary flow rate. K+ excretion rises in parallel with flow rate. Hence, the stimulatory effect of aldosterone on cellular K+ transport may be masked by opposite effects on tubular flow. Indeed, this was the case in a study combining clearance and microperfusion experiments.[110] Clearance experiments failed to reveal any significant kaliuresis upon acute infusion of aldosterone, whereas the same treatment resulted in marked K+ secretion in microperfused tubules. This apparent discrepancy was explained by a reduction of urinary flow rate following aldosterone-induced enhancement of tubular Na+ reabsorption.[110] Accordingly, Fimognari[13] observed aldosterone-dependent kaliuresis only in K+-depleted animals, i.e., under conditions when the flow dependence of K+ excretion was virtually lost.[111]

Thus, enhanced K+ secretion after the acute application of aldosterone is likely to result from the combined effect of (1) increased basolateral uptake via stimulation of the Na+-K+-ATPase (see above); (2) an increased electrochemical gradient favoring K+ movement across the apical membrane from the cell into the lumen; and (3) increased K+ conductance of the apical membrane.

A. GENOMIC VS. NONGENOMIC EFFECTS OF ALDOSTERONE

Clearance studies in the rat kidney[13] revealed differential effects of aldosterone on Na+ and K+ transport. Stimulation of K+ transport was independent of protein synthesis, which is consistent with a nongenomic action of aldosterone. In contrast, in the amphibian kidney, stimulation of the Na+/H+ exchanger by aldosterone was inhibited by spironolacton,[42] pointing to a genomic effect of the hormone. In this scenario, genomic activation of the exchanger was followed by nongenomic alkalinization of the cytoplasm, the trigger for opening of K^+ channels. Presently, it would be highly speculative to assume a nongenomic action of aldosterone on the renal Na^+/H^+ exchanger. However, given the extremely short latent period for activating this antiporter,[42,44,45] such speculations are tempting.

IV. EARLY EFFECTS OF ALDOSTERONE ON H^+ TRANSPORT

Aldosterone is involved in the regulation of acid-base homeostasis since both aldosterone deficiency and aldosteronism are accompanied by acidosis and alkalosis, respectively.[112] In the CCD where acid-base transporting intercalated cells are intermingled with Na^+ reabsorbing and K^+ secreting principal cells,[113] parts of H^+ secretion are related to rheogenic Na^+ transport which generates a lumen-negative transepithelial potential[17] and thereby facilitates H^+ secretion. Accordingly, this "electrically coupled" component of H^+ secretion will follow the same time course as Na^+ reabsorption.

However, the contributions of transport systems which are specifically regulated by aldosterone are of greater importance. Their response to aldosterone can already be seen after short latent periods. In turtle urinary bladder, stimulation of H^+ secretion preceded stimulation of Na^+ transport by 2 h and was observed as early as 1 h after addition of aldosterone.[114] This effect was suggested to be mediated by an increase in the capacity of the H^+ transporting systems, possibly caused by the formation of new pump sites. In papillary collecting ducts of adrenalectomized rats, acute administration of aldosterone restored the normal pH profile of this tubule segment.[115] In rabbit medullary collecting duct (MCD), tubular acidification was studied by determining the rate of reabsorptive HCO_3^- flux. It was significantly stimulated by aldosterone within 3 h regardless of whether amiloride was present in the luminal fluid or not.[116] In contrast, in frog diluting segment, aldosterone-dependent acidification of the luminal fluid was apparent 20 min after hormone application and was almost completely abolished by amiloride[117] consistent with the activation of plasma membrane Na^+/H^+ exchange. Hence, in frog diluting segment, aldosterone elicits a parallel increase in K^+ and H^+ secretion.[42,118]

Are other transporters involved in the aldosterone-induced rise of urinary acidification? In the apical membrane of MDCK cells, aldosterone stimulated an omeprazole-sensitive H^+/K^+ pump within 2 h.[119] In CCDs and MCDs from adrenalectomized rats, aldosterone augmented the H^+ pump activity within 3 h.[120] The effect could be seen both *in vivo* and *in vitro. In vitro* the earliest response was already visible after 30 min and activation strictly depended on protein synthesis. Since collecting tubule H^+-ATPases are known to undergo a cycle of endo- and exocytosis,[121,122] Khadouri[120] speculated whether aldosterone also induced exocytosis of preformed pump units.

This issue was approached directly by measuring cell membrane capacitances in mitochondria-rich cells of frog skin in response to aldosterone.[122] Cell membrane capacitance is a precise measure for the cell surface. Exocytosis-related increases of cell surface will thus be reflected by a rise in cell membrane capacitance. Since capacitances are measured in the whole-cell configuration of the patch clamp technique, one can also simultaneously assess membrane conductance.[123] Using this technique, it could be demonstrated that aldosterone caused a simultaneous rise in cell membrane capacitance and in H^+ pump current with a half-time of 12 min (Figure 4). Application of dicyclohexyl-carbodiimide (DCCD), a specific inhibitor of the H^+ pump reduced cell membrane conductance, but not cell surface. These results were taken as an indication that aldosterone promoted exocytotic insertion of H^+ pumps into the apical membrane of mitochondria-rich cells of frog skin.[122]

These studies convincingly demonstrated that several of the transporters relevant for acid-base homeostasis are under the control of aldosterone and were responsible for hormone-dependent acidification of the urine. However, urine pH per se is not a good indicator of whether total acid excretion was also stimulated by aldosterone, since it does not take into account the actual buffer capacity of the urine. Hence, urine pH may be very acidic but total acid elimination may only be marginal. This is exactly what was found in a clearance study investigating the effects of acute (2 h before urine collection) application of mineralo- and glucocorticoids on renal acid excretion.[124] Aldosterone caused a significant reduction of urine pH, but it had only minor effects on urinary titratable acid and it did not change urinary ammonium excretion. Thus, in these experiments, the overall contribution of aldosterone to acid elimination was not very pronounced.

The studies cited in this section provide indirect pharmacological evidence for a genomic action of aldosterone on H^+ transport. Hormone-dependent acid transport depended on protein synthesis,[120] or was related to the occupation of spironolacton sensitive aldosterone receptors.[42,114,117]

V. CONCLUSION

In this review we focused on the acute actions of aldosterone on Na^+, K^+ and H^+ transport. Aldosterone-dependent induction of these transport processes does not occur simultaneously, but in a sequence which is possibly related to the specific action of the hormone. Activation of the Na^+/H^+ exchanger is one of the earliest cellular responses to aldosterone. Several groups reported that it was apparent after a latent period of 10 to 20 min. The ensuing intracellular alkalinization had an immediate impact on pH-sensitive K^+ and Na^+ conductances and possibly on Na^+-K^+-ATPase activity. Transcriptional activity of Na^+-K^+-ATPase subunits, and as of yet unidentified proteins,[125] were also triggered after similar lag periods. Stated differently, all these events occurred before transepithelial Na^+ transport was stimulated.

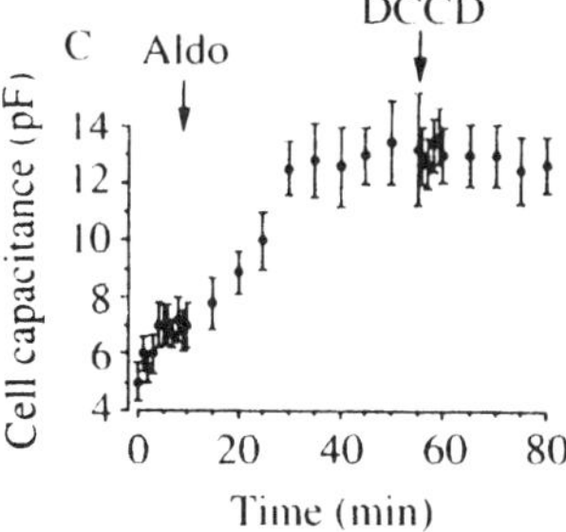

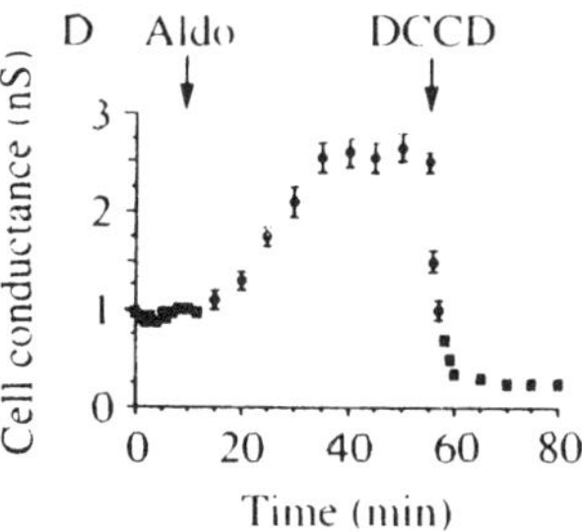

Figure 4:   Membrane capacitance and membrane conductance of mitochondria-rich cells from frog skin. Aldosterone simultaneously increases capacitance and conductance. The H$^+$-ATPase blocker dicyclohexylcarbodiimide (DCCD) only reduced membrane conductance without affecting membrane capacitance. This finding points to the insertion of H$^+$-ATPase containing vesicles. (From Harvey, B. J., *J. Exp. Biol.*, 172, 289, 1992. With permission.)

Hormone-dependent stimulation of transepithelial Na$^+$ transport was only apparent after 1 to 2 h. In many studies it was shown that it strictly depended, in contrast to aldosterone-dependent K$^+$ transport, on *de novo* protein synthesis. On the other hand, convincing evidence was provided that Na$^+$ channels, or the Na$^+$-K$^+$-ATPase, were not synthesized within this time, i.e., early activation of Na$^+$ transport demanded the *de novo* synthesis of regulatory proteins. The identification of these proteins and their mechanism of action will be one of the major challenges for future research in this field.

Identification of these proteins will also shed light onto the issue of genomic vs. nongenomic action of aldosterone. By which mechanism, for example, does aldosterone elicit the very rapid activation of the Na$^+$/H$^+$ exchanger? Does aldosterone bind to a plasma membrane receptor and act in a nongenomic way, or is the activation of the Na$^+$/H$^+$ exchanger the end point of the classical genomic signaling cascade of aldosterone?

It is tempting to speculate that intracellular pH might be a "second messenger" of aldosterone that does more than regulate ion transport across cell membranes. It is well known that intracellular alkalinization, brought about by stimulation of the Na$^+$/H$^+$ exchanger, is one of the earliest events following activation of quiescent cells and that it precedes metabolic activation. Cells lacking antiporter activity do not enhance DNA and, consequently, protein synthesis upon stimulation.[126] When measuring intracellular pH, a striking similarity was found between the response of Drosophila salivary gland cells to the steroid hormone ecdysone and the response of renal cells to aldosterone.[127] Both cell types rapidly alkalinized with almost identical time courses (10 to 20 min). In addition, in Drosophila salivary gland cells, gene activity was assessed by measuring nuclear volume. Alkalinization preceded nuclear swelling, which occurred after a 1 h lag period. Nuclear swelling could be prevented by inhibiting the Na$^+$/H$^+$ exchanger with amiloride.[127] It has yet to be established whether such a mechanism also holds true for the action of aldosterone:

aldosterone initially activates the Na^+/H^+ exchanger and the ensuing intracellular alkalinization is the trigger for the transcription of genes which are clearly necessary to elicit the early physiological response to the hormone. But again, this hypothesis still lacks experimental confirmation in target cells for aldosterone.

Another issue requires clarification. Does aldosterone exclusively affect transcellular transport — the topic of our review — or does aldosterone also rapidly regulate the permeability of the paracellular shunt pathway, and if it does so, what is the mechanism of action? Regulation of the paracellular conductance appears to be necessary for the following reason: in the CCD, aldosterone-dependent Na^+ reabsorption is accompanied by the generation of a lumen-negative transepithelial potential.[16] Hence, transcellular reabsorptive Na^+ transport is opposed by a back leak through the paracellular shunt pathway. Efficient Na^+ reabsorption would therefore require a decrease of the shunt conductance. To our knowledge, these theoretical considerations are supported only by few experimental studies looking at both long term[128] and short term effects[97,129] of aldosterone. Possibly, further improvement of the voltage-scanning technique[130] will provide more insight into this important question.

We want to finish this review by introducing a novel perspective in aldosterone action. Recently, we observed that the turnover of the nuclear pore complexes (NPCs) is regulated by aldosterone.[131] An individual NPC is a macromolecular complex composed of about 160 single proteins, anchored in the nuclear envelope and in charge of directed transport of inorganic ions and of macromolecules such as mRNAs and polymerases.[132] Thus, hormone-receptor complexes have to travel through a "central channel" located within the NPC to enter the nucleoplasm, and transcription products must exit through the same pathway to reach the ribosomes in the cytosol. Using atomic force microscopy,[133] we identified individual NPCs in the nuclear envelope of aldosterone-sensitive kidney cells in culture, and localized the central channel (Figure 5). Serum deprivation significantly reduced the number of NPCs per nucleus. This decrease did not occur when aldosterone was added to the serum-free medium. Although these experiments[131] were only performed 6 h after hormone supplementation, an "early response" to aldosterone in regulating NPC turnover is likely. It is attractive to speculate that aldosterone could also control the transport route that is responsible for signaling between cytosol and nucleoplasm.

ACKNOWLEDGMENTS

This article is dedicated to our friend and teacher, Prof. Gerhard Giebisch. Work cited from this laboratory was supported by Deutsche Forschungsgemeinschaft, SFB 176 (A6). We would like to thank L. D. T. Oriel GmbH, Darmstadt, FRG for providing us the Nanoscope III.

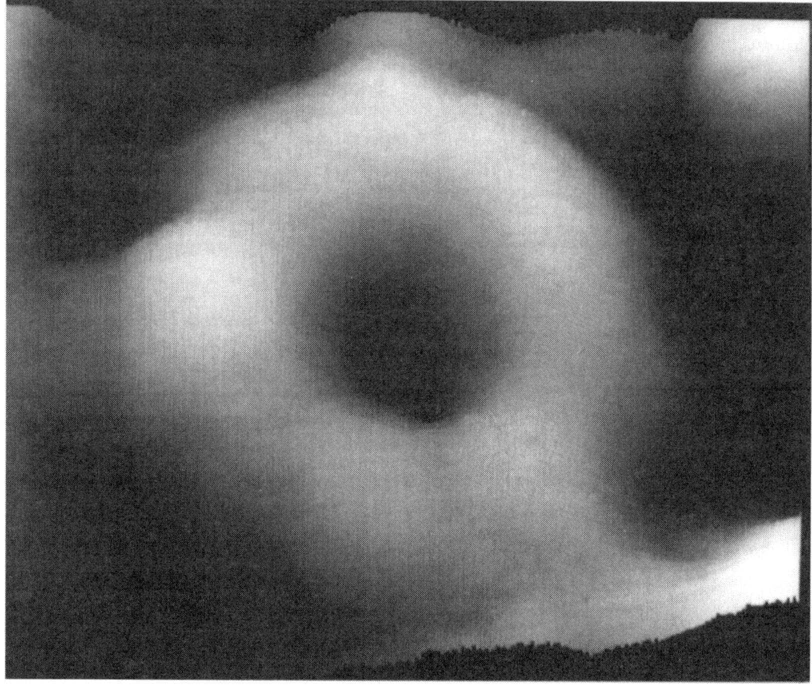

Figure 5: Single nuclear pore complex (NPC) of the nuclear envelope of a cultured kidney (MDCK) cell imaged by atomic force microscopy. The central channel is easily recognizable. Original width of the image: 200 nm.

REFERENCES

1. **Garty, H.,** Mechanisms of aldosterone action in tight epithelia, *J. Membr. Biol.,* 90, 193, 1986.
2. **Minuth, W. W., Steckelings, U., Gross, P.,** Complex physiological and biochemical action of aldosterone in toad urinary bladder and mammalian collecting duct cells, *Renal Physiol.,* 10, 297, 1987.
3. **Oberleithner, H.,** Acute aldosterone action in renal target cells, *Cell. Physiol. Biochem.,* 1, 2, 1991.
4. **Rossier, B. C., Palmer, L. G.,** Mechanisms of aldosterone action on sodium and potassium transport, in *The Kidney: Physiology and Pathopysiology,* 2nd ed., Seldin, D. W., Giebisch, G., Eds., Raven Press, New York, 1992, chap. 38.
5. **Funder, J. W.,** Aldosterone action, *Annu. Rev. Physiol.,* 55, 115, 1993.
6. **McKnight, A. D. C., DiBona, D. R., Leaf, A.,** Sodium transport across toad urinary bladder: a model "tight" epithelium, *Physiol. Rev.,* 60, 615, 1980.
7. **Civan, M. M., Garty, H.,** Toad urinary bladder as a model for studying transepithelial sodium transport, *Methods in Enzymol.,* 192, 683, 1990.
8. **Koefoed-Johnsen, V., Ussing, H. H.,** The nature of the frog skin potential, *Acta Physiol. Scand.,* 42, 298, 1958.
9. **Vandewalle, A., Farman, N., Bencsath, P., Bonvalet, J. P.,** Aldosterone binding along the rabbit nephron: an autoradiographic study on isolated tubules, *Am. J. Physiol.,* 240, F172, 1981.

10. **Lombès, M., Farman, N., Oblin, M. E., Baulieu, E. E., Bonvalet, J. P., Erlanger, B. F., Gasc, J. M.,** Immunohistochemical localization of renal mineralocorticoid receptor by using an anti-idiotypic antibody that is an internal image of aldosterone, *Proc. Natl. Acad. Sci. U.S.A.,* 87, 1086, 1990.

11. **Koeppen, B. M., Stanton, B. A.,** Sodium chloride transport. Distal nephron, in *The Kidney: Physiology and Pathophysiology,* 2nd ed., Seldin, D. W., Giebisch, G., Eds., Raven Press, New York, 1992, chap. 55.

12. **Ganong, W. F., Mulrow, P. J.,** Rate of change in sodium and potassium excretion after injection of aldosterone into the aorta and renal artery of the dog, *Am. J. Physiol.,* 195, 337, 1958.

13. **Fimognari, G. M., Fanestil, D. D., Edelman, I. S.,** Induction of RNA and protein synthesis in the action of aldosterone in the rat, *Am. J. Physiol.,* 213, 954, 1967.

14. **Horisberger, J.-D., Diezi, J.,** Effects of mineralocorticoids on Na^+ and K^+ excretion in the adrenalectomized rat, *Am. J. Physiol.,* 245, F89, 1983.

15. **Fanestil, D. D., Edelman, I. S.,** On the mechanism of action of aldosterone on sodium transport: effects of inhibitors of RNA and of protein synthesis, *Fed. Proc.,* 25, 912, 1966.

16. **Gross, J. B., Kokko, J. P.,** Effects of aldosterone and potassium-sparing diuretics on electrical potential differences across the distal nephron, *J. Clin. Invest.,* 59, 82, 1977.

17. **Gross, J. B., Imai, M., Kokko, J. P.,** A functional comparison of the cortical collecting tubule and the distal convoluted tubule, *J. Clin. Invest.,* 55, 1284, 1975.

18. **Crabbé, J.,** Site of action of aldosterone on the bladder of the toad, *Nature,* 200, 787, 1963.

19. **Lichtenstein, N. S., Leaf, A.,** Effect of amphotericine B on the permeability of the toad bladder, *J. Clin. Invest.,* 44, 1328, 1965.

20. **Sharp, G. W. G., Leaf, A.,** Mechanism of action of aldosterone, *Physiol. Rev.,* 46, 593, 1966.

21. **Nagel, W., Crabbé, J.,** Mechanism of action of aldosterone on active sodium transport across toad skin, *Pflügers Arch.,* 385, 181, 1980.

22. **Lindemann, B., Van Driessche, W.,** Sodium-specific membrane channels of frog skin are pores: current fluctuations reveal high turnover, *Science,* 195, 292, 1977.

23. **Palmer, L. G., Li, J. H.-Y., Lindemann, B., Edelman, I. S.,** Aldosterone control of the density of sodium channels in the toad urinary bladder, *J. Membr. Biol.,* 64, 91, 1982.

24. **Garty, H., Edelman, I. S.,** Amiloride-sensitive trypsinization of apical sodium channels, *J. Gen. Physiol.,* 81, 785, 1983.

25. **Palmer, L. G., Corthesy-Theulaz, I., Gaeggeler, H.-P., Kraehenbuhl, J.-P., Rossier, B.,** Expression of epithelial Na channels in Xenopus oocytes, *J. Gen. Physiol.,* 96, 23, 1990.

26. **Asher, C., Eren, R., Kahn, L., Yeger, O., Garty, H.,** Expression of the amiloride-blockable Na^+ channel by RNA from control versus aldosterone-stimulated tissue, *J. Biol. Chem.,* 267, 16061, 1992.

27. **Kleyman, T. R., Cragoe, E. J., Kraehenbuhl, J. P.,** The cellular pool of Na^+ channels in the amphibian cell line A6 is not altered by mineralocorticoids, *J. Biol. Chem.,* 264, 11995, 1989.

28. **Kleyman, T. R., Coupaye-Gerard, B., Ernst, S. A.,** Aldosterone does not alter apical cell-surface expression of epithelial Na^+ channels in the amphibian cell line A6, *J. Biol. Chem.,* 267, 9622, 1992.

29. **Tousson, A., Alley, C. D., Sorscher, E. J., Brinkley, B. R., Benos, D. J.,** Immunochemical localization of amiloride-sensitive sodium channels in sodium-transporting epithelia, *J. Cell Sci.,* 93, 349, 1989.

30. **Asher, C., Garty, H.,** Aldosterone increases the apical Na^+ permeability of toad bladder by two different mechanisms, *Proc. Natl. Acad. Sci. U.S.A.,* 85, 7413, 1988.

31. **Perkins, F. M., Handler, J. S.,** Transport properties of toad kidney epithelia in culture, *Am. J. Physiol.,* 241, C154, 1981.

32. **Hamill, O. P., Marty, A., Neher, E., Sakmann, B., Sigworth, F. J.,** Improved techniques for high-resolution current recording from cell-free membrane patches, *Pflügers Arch.,* 391, 85, 1981.

33. **Garty, H., Benos, D. J.,** Characteristics and regulatory mechanisms of the amiloride-blockable Na⁺ channel, *Physiol. Rev.,* 68, 309, 1988.

34. **Ling, B. N., Kemendy, A. E., Kokko, K. E., Hinton, C. F., Marunaka, Y., Eaton, D. C.,** Regulation of the amiloride-blockable sodium channel from epithelial tissue, *Mol. Cell. Biochem.,* 99, 141, 1990.

35. **Duchatelle, P., Ohara, A., Ling, B. N., Kemendy, A. E., Kokko, K. E., Matsumoto, P. S., Eaton D. C.,** Regulation of renal epithelial sodium channels, *Mol. Cell. Biochem.,* 114, 27, 1992.

36. **Schafer, J. A., Hawk, C. T.,** Regulation of Na⁺ channels in the cortical collecting duct by AVP and mineralocorticoids, *Kidney Int.,* 41, 255, 1992.

37. **Palmer, L. G.,** Epithelial Na channels: function and diversity, *Annu. Rev. Physiol.,* 54, 51, 1992.

38. **Horisberger, J.-D., Canessa, C., Rossier, B. C.,** The epithelial sodium channel: recent developments, *Cell Physiol. Biochem.,* 3, 283, 1993.

39. **Frindt, G., Sackin, H., Palmer, L. G.,** Whole-cell currents in rat cortical collecting tubule: low-Na diet increases amiloride-sensitive conductance, *Am. J. Physiol.,* 258, F562, 1990.

40. **Kemendy, A. E., Kleyman, T. R., Eaton, D. C.,** Aldosterone alters the open probability of amiloride-blockable sodium channels in A6 epithelia, *Am. J. Physiol.,* 263, C825, 1992.

41. **Pachà, J., Frindt, G., Antonian, L., Silver, R. B., Palmer, L. G.,** Regulation of Na channels of the rat cortical collecting tubule by aldosterone, *J. Gen. Physiol.,* 102, 25, 1993.

42. **Oberleithner, H., Weigt, M., Westphale, H.-J., Wang, W.,** Aldosterone activates Na⁺/H⁺ exchange and raises cytoplasmic pH in target cells of the amphibian kidney, *Proc. Natl. Acad. Sci. U.S.A.,* 84, 1464, 1987.

43. **Harvey, B. J., Ehrenfeld, J.,** Role of Na⁺/H⁺ exchange in the control of intracellular pH and cell membrane conductances in frog skin epithelium, *J. Gen. Physiol.,* 92, 793, 1988.

44. **Oberleithner, H., Kersting, U., Silbernagl, S., Steigner, W., Vogel, U.,** Fusion of cultured dog kidney (MDCK) cells: II. Relationship between cell pH and K⁺ conductance in response to aldosterone, *J. Membr. Biol.,* 111, 49, 1989.

45. **Vilella, S., Guerra, L., Helmle-Kolb, C., Murer, H.,** Aldosterone actions on basolateral Na⁺/H⁺ exchange in Madin-Darby canine kidney cells, *Pflügers Arch.,* 422, 9, 1992.

46. **Palmer, L. G., Frindt, G.,** Effects of cell Ca and pH on Na channels from rat cortical collecting tubule, *Am. J. Physiol.,* 253, F333, 1987.

47. **Wiesmann, W. P., Johnson, J. P., Miura, G. A., Chiang, P. K.,** Aldosterone-stimulated transmethylations are linked to sodium transport, *Am. J. Physiol.,* 248, F43, 1985.

48. **Minuth, W. W., Steckelings, U., Gross, P.,** Methylation of cytosolic proteins may be a possible biochemical pathway of early aldosterone action in cultured renal collecting duct cells, *Differentiation,* 36, 23, 1987.

49. **Sariban-Sohraby, S., Burg, M., Wiesmann, W. P., Chiang, P. K., Johnson, J. P.,** Methylation increases sodium transport into A6 apical membrane vesicles: possible mode of aldosterone action, *Science,* 225, 745, 1984.

50. **Kemendy, A. E., Eaton, D. C.,** Aldosterone-induced sodium transport in A6 epithelia is blocked by 3-deazaadenosine, a methylation blocker, *FASEB J.,* 4, A445, 1990.

51. **Chen, L., Williams, S. K., Schafer, J. A.,** Differences in synergistic actions of vasopressin and deoxycorticosterone in rat and rabbit CCD, *Am. J. Physiol.,* 259, F147, 1990.

52. **Girardet, M., Geering, K., Gaeggeler, H. P., Rossier, B. C.,** Control of transepithelial Na⁺ transport and Na-K-ATPase by oxytocin and aldosterone, *Am. J. Physiol.,* 251, F662, 1986.

53. **Oh, Y., Smith, P. R., Bradford, A. L., Keeton, D., Benos, D. J.,** Regulation by phosphorylation of purified epithelial Na⁺ channels in planar lipid bilayers, *Am. J. Physiol.,* 265, C85, 1993.

54. **Smith, P. R,. Benos, D. J.,** Epithelial Na⁺ channels, *Annu. Rev. Physiol.,* 53, 509, 1991.

55. **El Mernissi, G., Barlet-Bas, C., Khadouri, C., Cheval, L., Marsy, S., Doucet, A.,** Short-term effect of aldosterone on vasopressin-sensitive adenylate cyclase in rat collecting tubule, *Am. J. Physiol.,* 264, F821, 1993.

56. **Lipton, P., Edelman, I. S.,** Effects of aldosterone and vasopressin on electrolytes of toad bladder epithelial cells, *Am. J. Physiol.*, 221, 733, 1971.

57. **Eaton, D. C.,** Intracellular sodium ion activity and sodium transport in rabbit urinary bladder, *J. Physiol.*, 316, 527, 1981.

58. **Rick, R., Spancken, G., Dörge, A.,** Differential effects of aldosterone and ADH on intracellular electrolytes in toad urinary bladder epithelium, *J. Membr. Biol.*, 101, 275, 1988.

59. **Schmidt, U., Schmid, J., Schmid, H., Dubach, U. C.,** Sodium- and potassium-activated ATPase, *J. Clin. Invest.*, 55, 655, 1975.

60. **Horster, M., Schmid, H., Schmidt, U.,** Aldosterone *in vitro* restores nephron Na-K-ATPase of distal segments from adrenalectomized rabbits, *Pflügers Arch.*, 384, 203, 1980.

61. **Petty, K. J., Kokko, J. P., Marver, D.,** Secondary effects of aldosterone on Na$^+$-K$^+$-ATPase activity in the rabbit cortical collecting tubule, *J. Clin. Invest.*, 68, 1514, 1981.

62. **El Mernissi, G., Doucet, A.,** Short-term effect of aldosterone on renal sodium transport and tubular Na-K-ATPase, *Pflügers Arch.*, 399, 139, 1983.

63. **El Mernissi, G., Doucet, A.,** Short-term effect of aldosterone and dexamethasone on Na-K-ATPase along the rabbit nephron, *Pflügers Arch.*, 399, 147, 1983.

64. **El Mernissi, G., Doucet, A.,** Specific activity of Na-K-ATPase after adrenalectomy and hormone replacement along the rabbit nephron, *Pflügers Arch.*, 402, 258, 1984.

65. **Doucet, A., Katz, A. I.,** Short-term effect of aldosterone on Na-K-ATPase in single nephron segments, *Am. J. Physiol.*, 241, F273, 1981.

66. **Mujais, S. K., Chekal, M. A., Jones, W. J., Hayslett, J. P., Katz, A. I.,** Modulation of renal sodium-potassium-adenosine triphophatase by aldosterone, *J. Clin. Invest.*, 76, 170, 1985.

67. **Mujais, S. K., Chekal, M. A., Lee, S.-M. K., Katz, A. I.,** Relationship between adrenal steroids and Na-K-ATPase, *Pflügers Arch.*, 402, 48, 1984.

68. **Hayhurst, R. A., O'Neil, R. G.,** Time dependent actions of aldosterone and amiloride on Na$^+$-K$^+$-ATPase of cortical collecting duct, *Am. J. Physiol.*, 254, F689, 1988.

69. **Doucet, A.,** Function and control of Na-K-ATPase in single nephron segments of the mammalian kidney, *Kidney Int.*, 34, 749, 1988.

70. **Barlet-Bas, C., Khadouri, C., Marsy, S., Doucet, A.,** Enhanced intracellular sodium concentration in kidney cells recruits a latent pool of Na-K-ATPase whose size is modulated by corticosteroids, *J. Biol. Chem.*, 265, 7799, 1990.

71. **Blot-Chabaud, M., Wanstok, F., Bonvalet, J. P., Farman, N.,** Cell sodium-induced recruitment of Na$^+$-K$^+$-ATPase pumps in rabbit cortical collecting tubules is aldosterone-dependent, *J. Biol. Chem.*, 265, 11676, 1990.

72. **Coutry, N., Blot-Chabaud, M., Mateo, P., Bonvalet, J. P., Farman, N.,** Time course of sodium-induced Na$^+$-K$^+$-ATPase recruitment in rabbit cortical collecting tubule, *Am. J. Physiol.*, 263, C61, 1992.

73. **Pellanda, A. M., Gaeggeler, H. P., Horisberger, J. D., Rossier, B. C.,** Sodium-independent effect of aldosterone on initial rate of ouabain binding in A6 cells, *Am. J. Physiol.*, 262, C899, 1992.

74. **Barlet-Bas, C., Khadouri, C., Marsy, S., Doucet, A.,** Sodium-independent *in vitro* induction of Na$^+$, K$^+$-ATPase by aldosterone in renal target cells: permissive effects of triiodothyronine, *Proc. Natl. Acad. Sci. U.S.A.*, 85, 1707, 1988.

75. **Fujii, Y., Takemoto, F., Katz, A. I.,** Early effects of aldosterone on Na-K pump in rat cortical collecting tubules, *Am. J. Physiol.*, 28, F40, 1990.

76. **Eaton, D. C., Hamilton, K. L., Johnson, K. E.,** Intracellular acidosis blocks the basolateral Na-K pump in rabbit urinary bladder, *Am. J. Physiol.*, 247, F946, 1984.

77. **Schäfer, C., Westphale, H.-J., Oberleithner, H.,** Contrasting action of H$^+$ and Ca^{2+} on K$^+$ transport in the diluting segment of frog kidney, *Cell. Physiol. Biochem.*, 1, 286, 1991.

78. **Verrey, F., Kraehenbuhl, J. P., Rossier, B. C.,** Aldosterone induces a rapid increase in the rate of Na$^+$-K$^+$-ATPase gene transcription in cultured kidney cells, *Mol. Endocrinol.*, 3, 1369, 1989.

79. **Verrey, F., Schaerer, E., Zoerkler, P., Paccolat, M. P., Geering, K., Kraehenbuhl, J. P., Rossier, B. C.,** Regulation by aldosterone of Na^+,K^+-ATPase mRNAs, protein synthesis, and sodium transport in cultured kidney cells, *J. Cell Biol.,* 104, 1231, 1987.

80. **Geering, K., Giradet, M., Bron, C., Kraehenbuhl, J. P., Rossier, B. C.,** Hormonal regulation of (Na^+, K^+)-ATPase biosynthesis in the toad bladder, *J. Biol. Chem.,* 257, 10343, 1982.

81. **Geering, K., Claire, M., Gaeggeler, H.-P., Rossier, B. C.,** Receptor occupancy vs. induction of Na^+-K^+-ATPase and Na^+ transport by aldosterone, *Am. J. Physiol.,* 248, C102, 1985.

82. **Shahedi, M., Laborde, K., Bussières, L., Sachs, C.,** Acute and early effects of aldosterone on Na-K-ATPase activity in Madin-Darby canine kidney epithelial cells, *Am. J. Physiol.,* 264, F1021, 1993.

83. **Sharp, G. W. G., Leaf, A.,** Metabolic requirements for active sodium transport stimulated by aldosterone, *J. Biol. Chem.,* 240, 4816, 1965.

84. **Spooner, P. M., Edelman, I. S.,** Further studies on the effect of aldosterone on electrical resistance of toad bladder, *Biochim. Biophys. Acta,* 406, 304, 1975.

85. **Kirsten, E., Kirsten, R.,** Increased activity of enzymes of the tricarboxylic acid cycle in response to aldosterone in the toad bladder, *Pflügers Arch.,* 300, 213, 1968.

86. **Marver, D., Schwartz, M. J.,** Identification of mineralocorticoid target sites in the isolated rabbit cortical nephron, *Proc. Natl. Acad. Sci. U.S.A.,* 77, 3672, 1980.

87. **Minuth, W. W., Struck, M., Zwanzig, M., Gross, P.,** Action of aldosterone on citrate synthase in cultured renal collecting duct cells, *Renal Physiol. Biochem.,* 12, 85, 1989.

88. **Cortas, N., Abras, E., Arnaout, M., Mooradian, A., Muakasah, S.,** Energetics of sodium transport of the toad. Effect of aldosterone and sodium cyanide, *J. Clin. Invest.,* 73, 46, 1984.

89. **Garty, H., Edelman, I. S., Lindemann, B.,** Metabolic regulation of apical sodium permeability in toad urinary bladder in the presence and absence of aldosterone, *J. Membr. Biol.,* 74, 15, 1983.

90. **Silver, R. B., Frindt, G., Windhager, E. E., Palmer, L. G.,** Feedback regulation of Na channels in rat CCT. I. Effects of inhibition of Na pump, *Am. J. Physiol.,* 264, F557, 1993.

91. **Johnson, J. P., Green, S. W.,** Aldosterone stimulates Na^+ transport without affecting citrate synthase activity in cultured cells, *Biochim. Biophys. Acta,* 647, 293, 1981.

92. **Schmidt, T. J., Husted, R. F., Stokes, J. B.,** Steroid hormone stimulation of Na^+ transport in A6 cells is mediated via glucocorticoid receptors, *Am. J. Physiol.,* 264, C875, 1993.

93. **Adam, W. R., Ellis, A. G., Adams, B. A.,** Aldosterone is a physiologically significant kaliuretic hormone, *Am. J. Physiol.,,* 252, F1048, 1987.

94. **Lang, F., Rehwald, W.,** Potassium channels in renal epithelial transport regulation, *Physiol. Rev.,* 72, 1, 1992.

95. **Wright, F. S., Giebisch, G.,** Regulation of potassium excretion, in *The Kidney: Physiology and Pathophysiology,* 2nd ed., Seldin, D. W., Giebisch, G., Eds., Raven Press, New York, 1992, chap. 61.

96. **Koeppen, B. M., Biagi, B. A., Giebisch, G.,** Intracellular microelectrode characterization of the rabbit cortical collecting duct, *Am. J. Physiol.,* 244, F35, 1983.

97. **Horisberger, J.-D.,** Early effects of aldosterone on apical and basolateral membrane conductances of TBM cells, *Am. J. Physiol.,* 263, C384, 1992.

98. **Schafer, A., Troutman, S. L., Schlatter, E.,** Vasopressin and mineralocorticoid increase apical membrane driving force for K^+ secretion in rat CCD, *Am. J. Physiol.,* 258, F199, 1990.

99. **Wang, W., Giebisch, G.,** Dual effect of adenosine triphosphate on the apical small conductance K^+ channel of the rat cortical collecting duct, *J. Gen. Physiol.,* 98, 35, 1991.

100. **Wang, W., Schwab, A., Giebisch, G.,** Regulation of small-conductance K^+ channel in apical membrane of rat cortical collecting tubule, *Am. J. Physiol.,* 259, F494, 1990.

101. **Wiederholt, M., Schoormanns, W., Fischer, F., Behn, C.,** Mechanism of action of aldosterone on potassium transfer in the rat kidney, *Pflügers Arch.,* 345, 159, 1973.

102. **Stokes, J. B.,** Mineralocorticoid effect on K^+ permeability of the rabbit cortical collecting tubule, *Kidney Int.,* 28, 640, 1985.

103. **Hurst, A. M., Hunter, M.,** Apical K^+ channels of frog diluting segment: inhibition by acidification, *Pflügers Arch.,* 415, 115, 1989.

104. **Hurst, A. M., Hunter, M.,** Acute changes in channel density of amphibian diluting segment, *Am. J. Physiol.,* 259, C1005, 1990.

105. **Oberleithner, H., Guggino, W. B., Giebisch, G.,** Potassium transport in the early distal tubule of Amphiuma kidney: effects of potassium adaptation, *Pflügers Arch.,* 396, 185, 1983.

106. **Oberleithner, H., Lang, F., Wang, W., Messner, G., Deetjen, P.,** Evidence for an amiloride sensitive Na^+ pathway in the amphibian diluting segment induced by K^+ adaptation, *Pflügers Arch.,* 399, 166, 1983.

107. **Oberleithner, H., Dietl, P., Münich, G., Weigt, M., Schwab, A.,** Relationship between luminal Na^+/H^+ exchange and luminal K^+ conductance in diluting segment of frog kidney, *Pflügers Arch.,* 405 (Suppl.), S110, 1985.

108. **Harvey, B. J.,** Cellular mechanism of regulation of ion and water channels and pumps in high-resistance epithelia, in *Isotonic transport in leaky epithelia. Alfred Benzoa Symposium 34,* Ussing, H. H., Fischberg, I., Sten-Knudsen, O., Larsen, E. H., Willumsen, N. I., Eds., Munksgaard, Copenhagen, 1993, 312.

109. **Tsuchiya, K., Wang, W., Giebisch, G., Welling, P. A.,** ATP is a coupling modulator of parallel Na,K-ATPase - K-channel activity in the renal proximal tubule, *Proc. Natl. Acad. Sci. U.S.A.,* 89, 6418, 1992.

110. **Field, M. J., Stanton, B. A., Giebisch, G.,** Differential acute effects of aldosterone, dexamethasone and hyperkalemia on distal tubular potassium secretion in the rat, *J. Clin. Invest.,* 74, 1792, 1984.

111. **Khuri, R. N., Wiederholt, M., Strieder, N., Giebisch, G.,** Effects of flow rate and potassium intake on distal tubular potassium transfer, *Am. J. Physiol.,* 228, 1249, 1975.

112. **Gennari, F. J., Maddox, D. A.,** Renal regulation of acid-base homeostasis. Integrated response, in *The Kidney: Physiology and Pathophysiology,* 2nd ed., Seldin, D. W., Giebisch, G., Eds., Raven Press, New York, 1992, chap. 78.

113. **Schuster, V. L.,** Function and regulation of collecting duct intercalated cells, *Annu. Rev. Physiol.,* 55, 267, 1993.

114. **Al-Awqati, Q., Norby, L. H., Mueller, A., Steinmetz, P. R.,** Characteristics of stimulation of H^+ transport in turtle urinary bladder, *J. Clin. Invest.,* 58, 351, 1976.

115. **Higashihara, E., Carter, N. W., Pucacco, L., Kokko, J. P.,** Aldosterone effects on papillary collecting duct pH profile of the rat, *Am. J. Physiol.,* 246, F725, 1984.

116. **Stone, D. K., Seldin, D. W., Kokko, J. P., Jacobson, H. R.,** Mineralocorticoid modulation of rabbit medullary collecting duct acidification. A sodium-independent effect, *J. Clin. Invest.,* 72, 77, 1983.

117. **Weigt, M., Dietl, P., Silbernagl, S., Oberleithner, H.,** Activation of luminal Na^+/H^+ exchange in distal nephron of frog kidney. An early response to aldosterone, *Pflügers Arch.,* 408, 609, 1987.

118. **Westphale, H.-J., Schäfer, C., Oberleithner, H.,** Aldosterone stimulates cellular K^+ uptake and K^+ release in the diluting segment of the frog kidney, *Cell. Physiol. Biochem.,* 1, 89, 1991.

119. **Oberleithner, H., Steigner, W., Silbernagl, S., Vogel, U., Gstraunthaler, G., Pfaller, W.,** Madin-Darby canine kidney cells. III. Aldosterone stimulates an apical H^+/K^+ pump, *Pflügers Arch.,* 416, 540, 1990.

120. **Khadouri, C., Marsy, S., Barlet-Bas, C., Doucet, A.,** Short-term effect of aldosterone on NEM-sensitive ATPase in rat collecting tubule, *Am. J. Physiol.,* 257, F177, 1989.

121. **Schwartz, G. J., Al-Awqati, Q.,** Carbon dioxide causes exocytosis of vesicles containing H^+ pumps in isolated perfused proximal and collecting tubules, *J. Clin. Invest.,* 75, 1638, 1985.

122. **Harvey, B. J.,** Energization of sodium absorption by the H$^+$-ATPase pump in mitochondria-rich cells of frog skin, *J. Exp. Biol.,* 172, 289, 1992.

123. **Lindau, M., Neher, E.,** Patch-clamp techniques for time-resolved capacitance measurements in single cells, *Pflügers Arch.,* 411, 137, 1988.

124. **Wilcox, C. S., Cemerikic, D. A., Giebisch, G.,** Differential effects of acute mineralo- and glucocorticosteroid administration on renal acid elimination, *Kidney Int.,* 21, 546, 1982.

125. **Zwanzig, M., Minuth, W. W., Gross, P.,** Action of aldosterone on cultured collecting duct cells during the latent period, *Renal Physiol. Biochem.,* 13, 285, 1990.

126. **Pouysségur, J., Franchi, A., L'Allmain, G., Paris, S.,** Cytoplasmic pH, a key determinant of growth factor-induced DNA synthesis in quiescent fibroblasts, *FEBS,* 190, 115, 1985.

127. **Wünsch, S., Schneider, S., Schwab, A., Oberleithner, H.,** 20-OH-ecdysone swells nuclear volume by alkalinization in salivary glands of Drosophila melanogaster, *Cell Tissue Res.,* 274, 145, 1993.

128. **Uhlich, E., Baldamus, C. A., Ullrich, K.,** The effect of aldosterone on sodium transport in the collecting ducts of the mammalian kidney, *Pflügers Arch.,* 308, 111, 1969.

129. **Hoffmann, B., Clauss, W.,** Time-dependent effects of aldosterone on sodium transport and cell membrane resistances in rabbit distal colon, *Pflügers Arch.,* 415, 156, 1989.

130. **Köckerling, A., Sorgenfrei, D., Fromm, M.,** Electrogenic Na$^+$ absorption of rat distal colon is confined to surface epithelium: a voltage-scanning study, *Am. J. Physiol.,* 264, C1285, 1993.

131. **Oberleithner, H., Brinckmann, E., Krohne, G., Schwab, A.,** Imaging nuclear pores of aldosterone sensitive kidney cells by atomic force microscopy, *Proc. Natl. Acad. Sci.,* U.S.A., in press.

132. **Hinshaw, J. E., Carragher, B. O., Milligan, R. A.,** Architecture and design of the nuclear pore complex, *Cell,* 69, 1133, 1992.

133. **Oberleithner, H., Schwab, A., Wang, W., Giebisch, G., Hume, F., Geibel, J.,** Living renal epithelial cells imaged by atomic force microscopy, *Nephron,* 66, 8, 1994.

134. **Hunter, M.,** Personal communication, 1993.

Chapter 4

RAPID STEROID ACTIONS IN THE BRAIN: A CRITIQUE OF GENOMIC AND NONGENOMIC MECHANISMS

Miles Orchinik and Bruce S. McEwen

TABLE OF CONTENTS

I. INTRODUCTION

Steroid hormones produced in the gonads and adrenal glands cross the blood-brain barrier and exert profound influences on brain function and behavior. A striking example of the power of steroids to modulate behavior is the onset of behavioral estrus in female rodents. Two days exposure to elevated estrogen, followed by a short exposure to progesterone, produces a massive reorganization of female rat behaviors necessary for reproductive success. The endocrinology underlying these robust behavioral changes has been studied extensively; yet we know relatively little about the cellular and molecular mechanisms underlying these, and other, steroid effects in the brain. Long-term effects of steroids on neuronal function and morphology appear to be mediated, at least in part, by the regulation of gene expression. Short-term actions of steroids may also involve genomic mechanisms in addition to neuromodulatory actions mediated at the level of neuronal membranes. The goal of this chapter is to provide a sense of the range and complexity of mechanisms at the cellular and molecular level that steroids may utilize in neural tissue, and to begin to understand how multiple mechanisms may be integrated in the modulation of behavioral processes.

II. GENOMIC VS. NONGENOMIC STEROID ACTIONS

In the classical model of steroid hormone action, steroids freely diffuse across plasma membranes and bind to cytosolic or nuclear receptors in target cells. These steroid receptors belong to a family of soluble proteins that bind steroid and thyroid hormones, vitamin D metabolites, and retinoids.[1,2] Activation of these receptors by the cognate ligand enables the receptor to bind to specific DNA sequences, the hormone response elements. In this manner, steroid receptors act as ligand-dependent transcription factors to regulate gene expression.[3-6] Recently, it has been discovered that the target genes for steroids are not restricted to those possessing traditional hormone response elements in their promoter regions. Nuclear steroid receptors may also bind to other DNA sequences, such as the AP-1 site.[7-9] The interaction of steroid receptors with nuclear transcription factors may also regulate the activity of these proto-oncogene products. Additional complexity enters the classical model, because progesterone and estrogen receptors in certain cells can be activated independently of their cognate ligands[10,11] following phosphorylation or activation of dopamine receptors. Through these various genomic mechanisms, steroid receptors direct the synthesis of proteins in neural tissue, positively or negatively regulating the levels of neurotransmitter receptors, neuropeptides, growth factors, G proteins, metabolic enzymes, and so on.[12-16]

Some neurons respond to steroids in a manner that is incompatible with intracellular receptor-induced changes in protein synthesis. For example, certain steroids can alter ionic conductance in isolated membrane patches within seconds of steroid application.[17] The rapidity of these responses, and the absence of nuclear machinery in the membrane preparation, preclude a direct genomic mechanism of action. There are in fact numerous reports of steroid effects on neuronal activity that occur with a latency of seconds to very few minutes. (For reviews see References 13,18–20.) These effects are likely to be transient and to quickly decline upon removal of steroid.

Other lines of evidence also suggest the existence of alternate mechanisms of action. Steroids covalently bound to an immobilizing molecule, such as bovine serum albumin, do not freely cross neuronal membranes. Nevertheless, such molecules can rapidly alter neuronal excitability, neurosecretion, or behavior.[21–25] Steroid effects can be seen in preparations that lack cell nuclei,[26,27] or in the presence of protein synthesis inhibitors.[28–30] These rapid actions of steroids are presumed to involve receptors in neuronal membranes rather than cytosolic or nuclear receptors, and in a few instances steroid binding sites in neuronal membranes have been characterized (see below).

The *in vitro* evidence demonstrates that steroids have the potential to modulate neuronal function through nuclear transduction mechanisms and through mechanisms in neuronal membranes. On an organismal level, it is likely that multiple signal transduction pathways for any given steroid are integrated. How does one go about distinguishing between mechanisms when evaluating steroid effects on brain function or behavior? There are no generally applicable criteria for differentiating between receptor mechanisms. In many cases, especially when responses occur with a latency of minutes rather than seconds, kinetic analysis alone may be insufficient to distinguish between nuclear and membrane-mediated mechanisms. Relatively rapid genomic responses to steroids have been reported, at least at the level of mRNA, e.g., within 10 min of progesterone treatment, a change in mRNA levels of the proto-oncogene c-myc is evident in the chick oviduct.[31] On the other hand, membrane-mediated effects of steroids may develop with a significant delay. For example, neuronal responses to progesterone may require 15′ to appear, presumably due to the necessity of converting progesterone to neuronally active structures (see below). Peak levels of $GABA_A$ receptor-active 3α-hydroxy-5α-pregnan-20-one (3α-OH-DHP) may be found in the brain hours following systemic injection of a bolus of progesterone.[32]

There are further confounding factors in trying to distinguish between steroid mechanisms. Some steroidogenic pathways are reversible; the potential products of steroid metabolism have to be considered. For example, it has been assumed that actions of the progesterone metabolite, 3α-OH-DHP, are due to the modulation of receptors for γ-aminobutyric acid (GABA), the major inhibitory transmitter in the brain (see V.A. below). However, 3α-OH-DHP has been shown to stimulate the transcriptional activity of chicken progesterone receptors, presumably following enzymatic oxidation to 5α-pregnane-3,20-dione (5α-DHP).[33] Furthermore, steroid-induced protein synthesis itself may not be sufficient to delimit a mechanism of action. Immediate neuronal responses to steroids may involve membrane receptors, but the activation of second messengers by these receptors could secondarily activate gene expression. Until selective pharmacological agents are discovered that can differentiate between intracellular receptors and membrane-associated receptor subtypes, elucidation of the molecular mechanisms mediating neurobehavioral actions of steroids will be fraught with uncertainty.

III. STEROID METABOLISM IN THE BRAIN

The major steroids secreted by the gonads and the adrenal glands — estradiol, progesterone, testosterone, glucocorticoids, and mineralocorticoids

— can elicit changes in neuronal function; they are "neuroactive steroids." These circulating steroids may be converted in the brain to metabolites possessing either enhanced or diminished efficacy as neuromodulators. This enzymatic conversion may be a crucial step in both traditional and alternative steroid mechanisms of action. For example, during development, the aromatization of plasma-borne testosterone to estradiol may be a requisite step for masculinization of the brain.[34] In some species, enzymatic conversion of androgen in neural tissue is high enough to significantly raise the blood levels of estrogen.[35] Since aromatase activity has been localized in synaptosomes,[36] as well as microsomes, it is possible that steroids are metabolized and released at synaptic terminals where they could modulate synaptic transmission.

The major enzymatic pathways for conversion of progesterone in the female rat brain include 5α-reductase, producing 5α-DHP from progesterone, and 3α-hydroxysteroid oxidoreductase, reducing 5α-DHP to 3α-OH-DHP, a potent modulator of $GABA_A$ receptors (see below). Both 3α–hydroxysteroid oxidoreductase and 5α-reductase show low substrate specificity and can act on androgens, glucocorticoids, and mineralocorticoids, as well as several progestins.[37,38] 3α-Hydroxysteroid oxidoreductase catalyzes a reversible reaction that may account for the apparent transcriptional activity of 3α-OH-DHP mentioned above.[33] However, 5α-DHP does not appear to activate progesterone receptors in all species.[33,39]

While the rodent brain has considerable capacity for metabolizing circulating gonadal or adrenal steroids, a biosynthetic pathway for *de novo* production of pregnenolone from cholesterol is found in the brain.[40–42] The sulfate ester of pregnenolone, a potent modulator of Ca^{2+} channels (see below), is also found in brain.[43] Enzymes are present for the metabolism of pregnenolone to progesterone and ring A-reduced pregnanes, or alternatively to corticosterone.[42] The steroids derived from *de novo* synthesis in the brain have been termed "neurosteroids."[41] In gonadally and adrenally intact animals, the major sources of "neurosteroids" are peripheral endocrine glands rather than the brain. However, detectable levels of "neurosteroids" persist in the brains of gonadectomized and adrenalectomized rats, and synthesis may be regulated by mitochondrial benzodiazepine receptors.[44] It will be interesting to determine if any biologically active steroids are produced exclusively by the brain.

It may be misleading to generalize from rat data to other species concerning the role of brain-derived steroids. There appear to be striking species differences in the major pathways for progesterone metabolism in neural tissue. In the primate brains examined, progesterone is converted almost exclusively to 20α-hydroxy-4-pregnen-3-one (20α-DHP) by 20α-hydroxysteroid oxidoreductase.[45,46] This metabolite, 20α-DHP, lacks the 3α-hydroxyl required on ring A-reduced pregnanes for potent modulatory action at $GABA_A$ receptors in the rat brain, as well as the 20-ketone that confers maximal activity.[47–49] Therefore, in spite of the evidence that metabolism of progesterone in neural tissue is important for progesterone action in the rat, it seems that the sole source of $GABA_A$ receptor-active progestins in primates may be the peripheral endocrine system.

IV. STEROID RECEPTORS
IN NEURONAL MEMBRANES

During the last few years, evidence for nontraditional, receptor-mediated steroid actions has been rapidly accumulating, but the characterization of the receptors in question has not kept pace. Part of the problem appears to stem from difficulties inherent in performing direct binding studies using radiolabeled steroids. A high nonspecific signal is often generated when attempting to identify the moderate- to low-affinity recognition sites in neuronal membranes using steroidal radioligands. Steroids, being amphipathic, have affinity for the amphipathic phospholipid bilayer of plasma membranes. One wonders why steroids passively diffuse through plasma membranes and into the aqueous cytosolic environment, as the classical model of steroid action proposes. There is, in fact, evidence that steroids may be actively transported across cell membranes against a concentration gradient.[50] Nevertheless, in the 1970s several groups provided evidence for steroid binding sites in plasma membranes of peripheral tissue.[51–54] Since that time, there have been numerous reports of cell surface recognition sites in peripheral tissues, reviewed in this volume and elsewhere.[18,55]

Towle and Sze[56] reported the specific binding of tritiated testosterone, estradiol, progesterone, and corticosterone to partially purified synaptic membranes from rat brains. The equilibrium dissociation constants (K_d) for the binding of the gonadal steroids is approximately 10 nM, while the K_d for [^{3}H]corticosterone binding to membranes is 120 nM. Other labs have reported similar low-affinity binding of [^{3}H]corticosterone to neuronal membranes.[57] The glucocorticoid binding sites in neuronal membranes appear to be developmentally regulated,[58] suggesting that these receptors have a physiological role. However, some researchers have questioned the physiological relevance of low-affinity glucocorticoid receptors in the brain. Given the concentrations of circulating corticosterone in stressed rats, a more perplexing question might be to explain how the classical, high-affinity intracellular corticoid receptors mediate neuronal responses to stress. The Type I, or mineralocorticoid intracellular receptors, have a K_d of 0.5 to 1.0 nM for corticosterone, while Type II, or glucocorticoid receptors, have a K_d of 2.5 to 5 nM for corticosterone.[59–61] In non-stressed rats, the levels of circulating corticosterone may reach 600 nM on a daily basis, whereas corticosterone concentrations in stressed rats may be 1500 nM[59] with some estimates nearly twice that high. Assuming that only 10% of the circulating corticosterone is free to diffuse across the blood-brain barrier, the rest being bound to plasma binding proteins, neurons would be exposed to at least 150 nM corticosterone during stress. According to the principle of mass action, both types of intracellular receptors would be largely occupied by substress levels of corticosterone. In contrast, 150 nM free corticosterone in the brain during stress should occupy 55% of low-affinity (K_d = 120 nM), membrane-bound receptors. Therefore, a 100 to 150 nM K_d glucocorticoid receptor in neuronal membranes could be well suited for responding to rapidly fluctuating corticosterone levels during episodes of stress.

Electron microscopy has provided further evidence for steroid receptors in, or associated with, neuronal membranes in the rat brain. Using a monoclonal antibody against the intracellular rat glucocorticoid receptor, immunoreactivity is found associated with plasma membranes in the cell bodies and processes of hippocampal and hypothalamic neurons.[62] Similarly, estrogen receptor immunoreactivity in the guinea pig hypothalamus is seen in dendrites and in axon terminals, using a monoclonal antibody to the intracellular estrogen receptor.[63] A similar distribution of progesterone receptor immunoreactivity is seen at the electron microscopic level.[63] Recently, a variant form of the estrogen receptor that lacks exon 4, including the portion of the molecule required for nuclear translocation, was found to be abundant in the rat brain.[64] It is not yet clear if this molecule binds estrogen, or whether the estrogen receptor antibodies used also recognize the shortened form of the receptor. However, these studies raise the possibility that classical steroid receptors, or closely related molecules, may serve cellular functions in addition to those relating to gene expression.

One corticosteroid receptor in neuronal membranes has been extensively characterized, a receptor in amphibian brains.[65-69] These studies indicated that [3H]corticosterone binding to recognition sites in neuronal membranes differs from binding to intracellular receptors in terms of pharmacological specificity as well as sensitivity to guanyl nucleotides, ions, and reducing agents. Therefore, if these receptor proteins are similar to nuclear receptors, association of the receptor with plasma membranes must confer upon the molecule very different characteristics. In contrast to the low-affinity corticosteroid receptors characterized by Towle and Sze,[56] these receptors in amphibian brain membranes are characterized by high-affinity binding ($K_d \leq 0.5$ nM) of [3H]corticosterone, (Figure 1) perhaps reflecting the fact that circulating levels of glucocorticoids are at least an order of magnitude lower in these amphibians than in rats.

Several steroids are potent modulators of GABA$_A$ receptors (see below). The putative high-affinity recognition sites for steroids on GABA$_A$ receptors have not been characterized in direct binding studies, presumably due to the high nonspecific (nondisplaceable) binding. Instead, low-affinity binding to presumed sites on GABA$_A$ receptors has been described for less hydrophobic, but less efficacious steroids. [3H]Pregnenolone sulfate binds in a complex manner to neuronal membranes with two or three dissociation constants ranging from 300 to 500 nM to 300 μM.[70] [3H]Dehydroepiandrosterone sulfate (DHEAS) was reported to bind to neuronal membranes in a biphasic manner, with $K_d = 3$ μM and 550 μM.[71,72] Interpretation of these data is difficult because low-affinity radioligand binding was quantified using a filtration assay, which is best suited for high-affinity radioligands. For a number of reasons, it is not clear whether these steroids are binding to GABA$_A$ receptors or intercalating into the lipid bilayer.

Another approach that has been used to identify binding sites in membranes involves immobilizing the steroid to reduce the possibility of radioligand

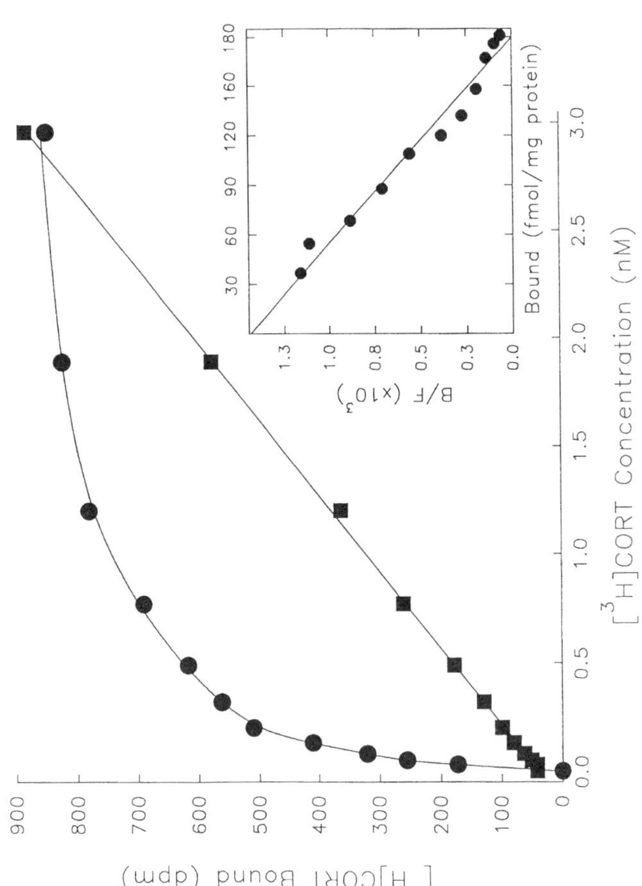

Figure 1. High-affinity binding of [³H]corticosterone to amphibian neuronal membranes: equilibrium saturation binding isotherm. Well washed crude synaptosomal membranes from whole brains of *Taricha granulosa* were incubated with the concentrations of [³H]CORT shown for 2 h at 30°C in buffer containing 10 nM MgCl$_2$. Specific binding (circles) equals total binding (not shown) minus nonspecific binding (squares). Data were best fit with a one-site model with K_d = 0.14 ± 0.01 nM and B_{max} = 183 ± 4 fmol/mg protein. Inset: Scatchard replot of specific binding data was linear; Hill coefficient = 1.08. (Modified from Orchinik, M., Murray, T.F., Franklin, P.H., and Moore, F.L., *Proc. Natl. Acad. Sci. U.S.A.*, 89, 3830, 1992.)

binding to intracellular receptors in the membrane preparation. An iodinated progesterone-bovine serum albumin complex, [^{125}I]BSA-11-P (progesterone conjugated at C11 to BSA), has been used to label putative binding sites for progesterone in neuronal membranes.[73,74] [^{125}I]BSA-11-P has no known biological activity,[75] binds to neuronal membranes with very low affinity (1600 to 1700 nM), and is not easily displaced by unlabeled progesterone, making its usefulness as a radioligand somewhat limited. However, BSA conjugated to progesterone at C3 (BSA-3-P), displaces [^{125}I]BSA-11-P binding with an estimated K_i of 30 nM, stimulates hypothalamic GnRH release,[23,75] alters α-1 adrenergic receptor-mediated cAMP production,[24] and facilitates female rat reproductive behavior[25] (see below). Interestingly, sex differences in [^{125}I]BSA-11-P binding were reported, and the binding of [^{125}I]BSA-11-P to membranes is sensitive to estradiol treatment.[74]

A potential problem in studying the binding of gonadal and adrenal steroids to plasma membranes is the presence of steroid binding globulins in the blood. We have found that [^{3}H]corticosterone binds with high affinity (K_d = 6.5 nM) to well-washed synaptosomal membranes prepared from prairie vole brains, but that this high-affinity binding is disrupted when the brains are saline perfused prior to decapitation.[76] The high-affinity binding is restored when plasma is added back to membranes prepared from saline-perfused brains. The plasma component required for high-affinity binding is likely to be corticosteroid binding globulin (CBG), since CBG and gonadal steroid binding proteins have been shown to bind to plasma membranes from peripheral tissues and stimulate adenylate cyclase activity.[77–79] It is unlikely that all steroid binding to neuronal membranes involves plasma-binding globulins, because some binding studies have used perfused brains, or determined that perfusion does not eliminate binding to neuronal membranes.[57,65]

V. ALTERNATE TRANSDUCTION MECHANISMS

A. MODULATION OF LIGAND-GATED ION CHANNELS

The ligand-gated ion channels comprise a superfamily of receptors that include GABA$_A$ receptors, nicotinic cholinergic receptors, strychnine-sensitive glycine receptors, and ionotropic glutamate receptors.[80] The most compelling evidence for steroid modulation of ligand-gated ion channels comes from studies on GABA$_A$ receptors. GABA$_A$ receptors are oligomeric complexes composed of multiple subunits that form an integral chloride channel. The receptors contain a number of distinct, allosterically interacting binding sites for GABA, benzodiazepines, barbiturates, and channel blockers. GABA$_A$ receptor function is also modulated by steroids; the naturally occurring metabolites of progesterone and deoxycorticosterone, 3α-OH-DHP and 3α,21-OH-5α-pregnan-20-one (5α-THDOC), respectively, are among the most potent GABA$_A$ receptor ligands known. (For reviews see References 17,47,55,81-85.) Nanomolar concentrations of these reduced pregnane steroids can enhance GABA-induced Cl$^-$ currents in neurons. In contrast, physiological concentrations of the major circulating hormonal steroids do not appear to directly modulate $GABA_A$ receptors (Figure 2). Since the steroid modulation of $GABA_A$ receptor function has been extensively reviewed, we will only mention a few recent developments.

Even in the absence of direct radioligand binding studies, there is considerable evidence that steroid recognition sites reside on the $GABA_A$ receptor, and not in the lipid bilayer surrounding the receptor. The transient expression of $GABA_A$ receptor subunits can confer steroid sensitivity to nonresponsive cells.[86–91] Since steroids applied intracellularly cannot mimic the actions of steroids applied extracellularly on GABA-sensitive chloride currents, the steroid binding site appears to be on the cell surface.[92] In addition, steroids can modulate ligand binding to solubilized $GABA_A$ receptors in a manner similar to membrane-bound receptors.[93]

The possibilities for $GABA_A$ receptor heterogeneity are enormous, since multiple isoforms for each of the subunits exist, and a heterogeneous population of $GABA_A$ receptors appears to be present in vertebrate brains.[94–98] An intriguing idea is that different combinations of receptor subunits form $GABA_A$ receptors with differing sensitivities to steroids. One group has found that steroids appear to modulate $GABA_A$ receptor function independently of subunit composition.[86] However, other studies suggest that the efficacy of steroids to modulate $GABA_A$ receptors is, at least in part, subunit-specific.[88,90,99] Consistent with the possibility that differences in subunit composition might confer regional differences in sensitivity to steroids, there are regional differences in steroid modulation of radioligand binding to $GABA_A$ receptors.[100–104] In addition, $GABA_A$-like autoreceptors are modulated by benzodiazepines and barbiturates, but not by steroids.[105] Therefore, regional differences in the steroid sensitivity of $GABA_A$ receptors appear to exist, and may reflect differences in receptor composition. If $GABA_A$ receptor subunit expression is regulated by experience or the hormonal environment, this might represent a locus at which steroids modulate brain function through both genomic and nongenomic mechanisms.

Considering the structural similarities between ligand-gated ion channels, it seems reasonable that steroids might modulate other ligand-gated ion channels as well. Several studies suggest that this might be the case. Low micromolar concentrations of corticosteroids can potentiate glycine-induced Cl^- currents,[106] whereas high micromolar concentrations of pregnenolone sulfate, deoxycorticosterone, and progesterone can inhibit glycine-induced Cl^- currents.[107] The different rank order potencies of steroids at the glycine and $GABA_A$ receptors suggest that the receptors have different structural requirements for steroid binding.

Some evidence indicates that the *N*-methyl-*D*-aspartate (NMDA) subtype of glutamate receptor may be sensitive to modulation by steroids. Pregnenolone sulfate potentiates both NMDA-induced Ca^{2+} currents[108] and NMDA-induced increases in intracellular Ca^{2+} concentration.[109] In addition, DHEAS can rapidly increase the excitability of CA1 neurons in the rat hippocampus in re-

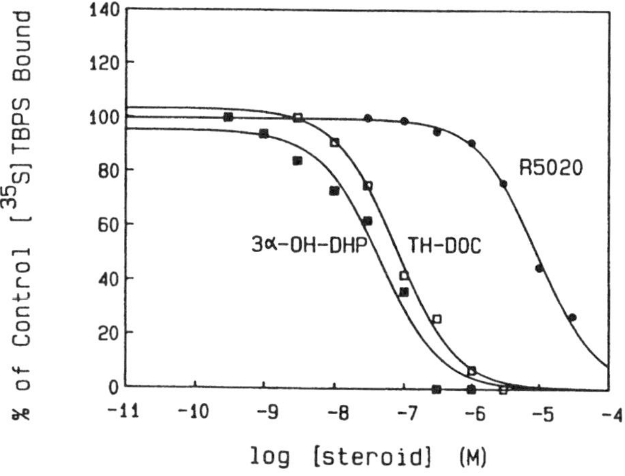

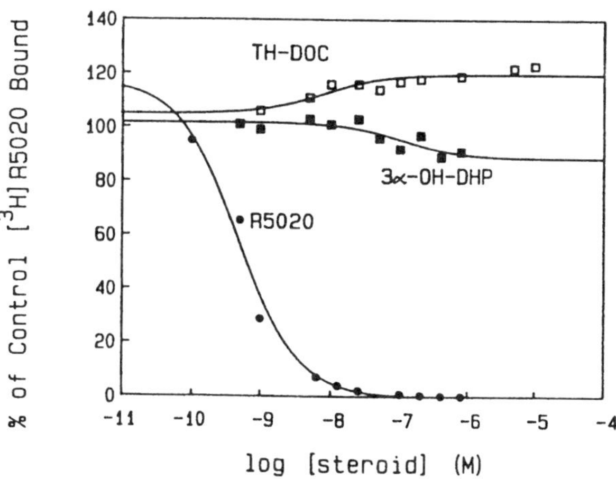

Figure 2. Effect of $GABA_A$ receptor-active steroids and R5020 (promegesterone) on [^{35}S]TBPS (upper panel) and [^{3}H]R5020 (lower panel) in rat cerebral cortex and uterus, respectively. TBPS (t-butylbicyclophosphorothionate) binds to a site on or near the $GABA_A$ receptor Cl$^-$ channel, whereas R5020 is a specific, high-affinity ligand for the cytosolic progesterone receptor. Cortical P2 membrane fractions or uterine cytosol were incubated with various concentrations of the steroids shown. 3α-OH-DHP and 5α-THDOC were potent inhibitors of [^{35}S]TBPS binding to $GABA_A$ receptors, but did not compete for [^{3}H]R5020 binding sites on the cytosolic progesterone receptor. In contrast, unlabeled R5020 was a potent inhibitor of [^{3}H]R5020 binding to progesterone receptors, but not a potent inhibitor of [^{35}S]TBPS binding. (From Gee, K.W., Bolger, M.B., Brinton, R.E., Coirini, H., and McEwen, B.S., *J. Pharmacol. Exp. Ther.*, 246, 803, 1988. With permission.)

sponse to Schaffer collateral stimulation,[110] a response that may involve NMDA receptors. The concentrations of pregnenolone sulfate and DHEAS used in these experiments was in the mid- to high-micromolar range.

The nicotinic cholinergic receptor is another ligand-gated ion channel that may be sensitive to modulation by steroids. Progesterone can depress acetylcholine-induced currents, with an IC_{50} of 7 to 9 μM, in cells expressing nicotinic cholinergic receptor subunits.[111,112] The response occurs in outside-out patches, so it is unlikely to depend on cytoplasmic components such as those related to the membrane receptors for progesterone in *Xenopus* oocytes (see below).

B. G PROTEIN-COUPLED RESPONSES

The potential for direct steroid modulation of ligand-gated ion channels, particularly the $GABA_A$ receptor chloride ionophore, has been well documented. Less well known are the numerous reports of rapid steroid effects on neuronal activity that appear to involve guanine nucleotide binding regulatory proteins (G proteins). Evidence for this type of mechanism includes steroid activation of G proteins, G protein-regulated second messenger systems, or ion channels, as well as blockade of steroid effects following inactivation of G proteins, and guanine nucleotide regulation of binding to membrane-bound steroid receptors.

In many cases, the concentrations of steroid required for *in vitro* modulation of ligand-gated ion channels are supraphysiological. In contrast, several steroids can modulate voltage-sensitive Ca^{2+} channels with nanomolar or subnanomolar potency. For example, low nanomolar concentrations of estradiol can suppress Ca^{2+} currents in striatal neurons within seconds of application.[113] Similarly, nanomolar concentrations of pregnenolone sulfate and allotetrahydrocorticosterone can inhibit Ca^{2+} currents in hippocampal CA1 neurons.[114,115] Several neurotransmitters can inhibit voltage-sensitive Ca^{2+} channels through G protein-dependent mechanisms,[116,117] and it appears that steroids may activate similar transduction pathways. The inhibition of Ca^{2+} currents by steroids in hippocampal and striatal neurons is sensitive to either guanyl nucleotides or G protein inactivation by ADP-ribosylation. This type of mechanism could account for the rapid modulation of dopamine release from striatal neurons by estradiol,[118] or a number of other rapid steroid effects on neurosecretion.[26–28,75,119–124]

Estradiol has been shown to alter neuronal excitability through other G protein-coupled mechanisms, the rapid modulation of a potassium conductance in amygdala, and hypothalamic neurons.[29] This estradiol effect appears to be mediated by cAMP,[125] consistent with an estradiol receptor/G protein activation of adenylate cyclase.

These studies are consistent with the existence of G protein-coupled steroid receptors, whereas other studies suggest a more complex interaction of steroids with G protein-mediated transduction pathways. It appears that steroids can rapidly alter the efficacy of coupling between certain neurotransmitter receptors and G proteins or effector mechanisms. For example, brief exposure of cerebellar slices to estradiol potentiates phosphatidylinositol hydrolysis in response to quisqualate.[126] In this case, estradiol appears to be altering the second messenger systems associated with metabotropic glutamate receptors that couple to phospholipase C activity via G proteins. This rapid potentiation of neurotransmitter-evoked phospholipase C activity could underlie the enhancement of quisqualate-induced neuronal excitability following estradiol administration.[126] Estradiol can rapidly potentiate neuronal responses to glutamate evoked through non-NMDA glutamate receptors in the hippocampus as well,[127] although it is not clear whether ionotropic or G protein-coupled metabotropic glutamate receptors are involved. Interestingly, the rapid response to estradiol occurs more frequently in neurons from estradiol-primed vs. ovariectomized animals, suggesting a convergence of long-term genomic and short-term membrane-mediated mechanisms in the hippocampus.

In addition to potentiating neurotransmitter receptor-stimulated G protein-mediated pathways, steroids can also inhibit neuronal responses to certain neurotransmitters. In brain slices, progesterone or BSA-3-P can inhibit norepinephrine-stimulated cAMP accumulation within 5 min.[24] More specifically, it appears that progesterone blocks the α1-receptor enhancement of β-adrenergic-receptor-stimulated adenylate cyclase activity. This effect depends on prior estradiol-priming, again suggesting an interaction of long-term and short-term steroid actions. Several neurotransmitter receptors, including μ opiate and $GABA_B$ receptors, modulate K^+ conductance through a G protein-mediated process. Estradiol has rapid[128] and delayed[129] effects on a μ-opiate and $GABA_B$ receptor-sensitive K^+ conductance in arcuate neurons. Within several minutes of exposure to estradiol, these neurons show a decreased responsiveness to μ opiate or $GABA_B$ agonists that is reflected in an attenuated transmitter-induced K^+ conductance.

How does one account for these phenomena mechanistically? One explanation is the existence of G protein-coupled steroid receptors in neuronal membranes similar to neurotransmitter receptors. The first evidence for G protein-coupled steroid receptors came from studies on the mechanisms mediating progesterone's effect on *Xenopus* oocyte maturation,[130–134] although there is also evidence suggestive of G protein-coupled steroid receptors in other peripheral tissue, such as human sperm[135,136] and rat pituitary.[137,138] Radioligand binding studies using neuronal membranes provide some evidence to support this hypothesis in the brain. Corticosteroid receptors in amphibian neuronal membranes are negatively modulated by guanyl nucleotides in a manner consistent with direct interaction of the steroid recognition site with G proteins (Figure 3).[66] [^{3}H]Corticosterone binding is inhibited by nonhydrolyzable guanyl nucleotides, an effect that appears to be due to the rapid dissociation of corticosterone from receptors that is probably induced by a negative heterotropic interaction between receptor and G protein.

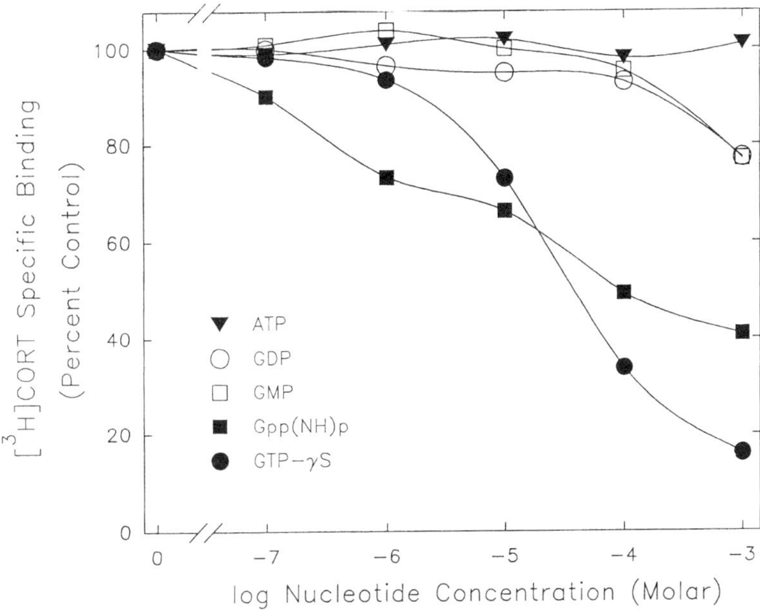

Figure 3. Effect of guanyl nucleotides on equilibrium binding of [³H]corticosterone to amphibian neuronal membranes. Well washed *Taricha* whole brain membranes were incubated with 0.75 nM [³H]CORT and the concentrations of nucleotide shown. Nonhydrolyzable guanyl nucleotides inhibited specific binding with IC$_{50}$ estimates of 43.2 ± 11 μM for GTP-γS and 104 ± 65 μM for Gpp(NH)p. The negative modulation of corticosterone binding by GTP analogs is consistent with the existence of G protein-coupled corticosteroid receptors in neuronal membranes. (Reproduced from Orchinik, M., Murray, T.F., Franklin, P.H., and Moore, F.L., *Proc. Natl. Acad. Sci. U.S.A.*, 89, 3830, 1992.)

Another possibility is that steroids act directly at the level of G protein to alter the interaction of G proteins with either the effector molecules or with neurotransmitter receptors. This possibility is supported by studies showing that several compounds can act as G protein "antagonists" by blocking the interaction of G proteins with neurotransmitter receptors[139,140] or as receptor-independent activators of G proteins.[141] There is currently no direct evidence demonstrating that steroids act in this manner, but such a mechanism might explain how steroids can rapidly alter the responsiveness of a given cell to several neurotransmitters or neuropeptides.

A steroid-induced change in neurotransmitter receptor-G protein coupling could be manifested in a change in affinity state of the receptors for agonist binding. This type of phenomenon may be seen in the rapid decrease in the fraction of hypothalamic muscarinic receptors showing high-affinity for agonist binding following estradiol treatment,[142] or the rapid estradiol-induced conversion of D2 dopamine receptors from high into low affinity states.[143] However, an estradiol treatment that altered K$^+$ conductance in response to μ opiate receptor activation did not alter the relative fraction of high affinity agonist binding sites for μ opiate receptors.[144]

While the molecular basis for these diverse steroid effects on G protein-mediated transduction pathways is clearly unsettled, overall the evidence indicates that steroids can rapidly alter brain function through membrane mechanisms that are independent of ligand-gated ion channels. Steroid modulation of signal transduction pathways in neuronal membranes involving G protein-regulated second messenger systems or ion channels could have significant effects on neuronal excitability or synaptic strength.

VI. FUNCTIONAL STUDIES

It is one thing to demonstrate that steroids can modulate ligand-gated ion channels or a G protein-coupled receptors *in vitro,* but quite another task to demonstrate that these transduction pathways are relevant to physiology or behavior. In this section, we will discuss three representative areas of research that point to the functional significance of nontraditional actions of steroids.

A. PROGESTERONE AND THE ONSET OF BEHAVIORAL ESTRUS

The onset of receptivity in female rodents is the best studied model for steroid regulation of behavior, involving the complex interplay of steroid hormones with numerous neurotransmitter and neuropeptide systems. During the estrus cycle, plasma concentrations of estradiol increase gradually over 2 d of diestrus and proestrus. The increase in estradiol is followed by a fivefold increase in plasma progesterone that occurs in 2 to 4 hours just prior to ovulation and the onset of behavioral estrus. This priming action of estradiol, followed by the rapid actions of progesterone, produces a profound reorientation of female behavior, such that the female displays receptive behavior at the time of ovulation. The prolonged exposure to estrogen may regulate the synthesis of numerous neuropeptides, neurotransmitters, and receptors in the hypothalamus.[145,146] One of the more robust estradiol effects is the induction of intracellular progesterone receptors that were assumed to mediate all of the actions of progesterone. However, there were difficulties because of the reports of short-latency effects of progesterone that occurred in brain regions with very few, if any, detectable progesterone receptors. This section will focus on some of the rapid effects of progesterone in facilitating lordosis behavior.

In many rodents, the ventromedial hypothalamus (VMH) is an essential site for the priming action of estradiol, as well as the lordosis-facilitating actions of progesterone. In some species, the midbrain has also been implicated in the lordosis-facilitating actions of progesterone. Electrophysiological studies performed on freely moving golden hamsters have provided insights into how exposure to progesterone alters neuronal activity during the onset of behavioral receptivity.[147–149] It appears that the actions of progesterone in estradiol-primed hamsters are not restricted to a small subset of neurons; rather, there is a change in excitability or a functional reconfiguration in a large majority of midbrain and hypothalamic neurons following progesterone injection. The functional reconfiguration includes changes in the responsiveness of neurons to somatosensory stimuli or to lordosis promoting- vs. lordosis-inhibiting movements. Consistent with the existence of multiple signal transduction pathways for progesterone, the latencies of the neuronal responses to progesterone injection vary widely. In both hypothalamus and midbrain, the activity level of some neurons is altered within 5 to 6 minutes of progesterone injection, whereas other neurons do not exhibit electrophysiological changes until 6 h postinjection. Overall, the responsiveness of neurons to progesterone is cumulative, corresponding to the onset of maximal behavioral responsiveness. Some of the progesterone-induced changes in neuronal activity are dependent upon prior exposure to estradiol, while other responses occur independently of estradiol priming. These studies demonstrate that the profound reorientation of behavior induced by progesterone is correlated with massive changes in neuronal activity and functional state. In addition, it appears that activation of multiple, summating mechanisms in several brain regions is required for the full expression of behavior.

Other studies support the conclusion that both long-term and short-term mechanisms for progesterone action are required for full display of lordosis behavior in hamsters. Receptive behavior can be fully induced in estradiol-primed animals by injecting progesterone into both the ventral tegmental area (VTA) of the midbrain, and the VMH. Of particular interest is the finding that the full expression of lordosis responding can be elicited following injection of P-3-BSA into the midbrain and free progesterone into the hypothalamus.[150] However, the converse treatment, P-3-BSA in the hypothalamus and free progesterone in the VTA, is behaviorally ineffective, suggesting that progesterone is acting at the level of neuronal membrane in the midbrain, but via intracellular receptors in the hypothalamus. The presumed membrane-mediated response in the ventral tegmental area is elicited by P-3-BSA, but not by P-11-BSA,[25] consistent with the efficacy of P-3-BSA, but not P-11-BSA, in mimicking the neurosecretory actions of progesterone.[75] Since a 3α-hydroxyl configuration on ring A-reduced pregnane steroids is required for activity at $GABA_A$ receptors,[47-49] it seems unlikely that P-3-BSA could be acting through $GABA_A$ receptors. Perhaps the $[^{125}I]$P-11-BSA binding site, described above,[73,74] is involved.

However, that may be a hasty conclusion. When the 5α-reduction of progesterone is blocked, inhibiting the formation of $GABA_A$ receptor-active metabolites, the behavioral response to progesterone injection into the VTA is blocked.[151] Also, $GABA_A$ receptor-active pregnanes applied to the VTA facilitate lordosis behavior.[152] These findings are difficult to reconcile with the induction of receptivity by P-3-BSA. Is P-3-BSA not subject to the apparently stringent structural requirements for interaction with $GABA_A$ receptors? Or alternatively, is P-3-BSA, but not P-11-BSA, cleaved in the VTA, not in the VMH, and the free progesterone metabolized to $GABA_A$-active compounds in the

VTA? Does progesterone have multiple pathways for membrane action in the VTA? These studies suggest that at least one proximal mechanism for action of progesterone in the midbrain is through modulation of $GABA_A$ receptors.

Can we conclude from these data that progesterone acts exclusively through genomic mechanisms in the hypothalamus and nongenomic mechanisms in the VTA? In the electrophysiological studies, changes in hypothalamic neuronal activity emerged within 5 to 6 min of progesterone injection. These responses are unlikely to be mediated by intracellular receptors, but we do not know whether these changes reflect direct actions of progesterone (or metabolites) on hypothalamic neuronal membranes, or indirect actions on neurons that project to the hypothalamus. In the site-specific injection studies and electrophysiological studies, there was a delay before the onset of maximal lordosis responsiveness. Therefore, it seems that rapid progesterone actions in the hypothalamus are insufficient for the induction of lordosis responses, but they may be a necessary component in the cumulative, integrated action of progesterone. Presumably some of the delayed actions of progesterone in the hypothalamus involve protein synthesis, but conceivably this could reflect activity-dependent changes in gene expression, rather than intracellular receptor-mediated transcriptional action. These studies underline the difficulty in describing the actions of progesterone as genomic or nongenomic.

In this discussion, we have focused on rapid actions of progesterone, rather than the actions of estradiol. In the rat, the priming action of estradiol may include regulating $GABA_A$ receptor density in the hypothalamus;[153–155] this might prime the brain to respond rapidly to increases in $GABA_A$ receptor-active progestins. Somewhat more perplexing are the rapid neurophysiological responses to estradiol,[29,156,157] since the behavioral responses to estradiol generally have a delayed onset. Part of the explanation may be that the membrane-mediated actions of estradiol are dependent upon prior treatment with estradiol. In the hippocampus at least, the priming action of estradiol enhances the electrophysiological responses to subsequent application of estradiol,[127] suggesting that actions of estradiol during diestrus and early proestrus may sensitize neurons to estradiol exposure through the induction of estradiol membrane receptors or estradiol-sensitive ion channels. Another interesting possibility is that the induction of estradiol receptors in neuronal membranes primes the brain to respond rapidly to the precipitous drop in estradiol levels that occurs just prior to estrus.

B. ESTRADIOL AND THE REGULATION OF HIPPOCAMPAL EXCITABILITY

Some of the best examples of the interplay between rapid and delayed effects of steroids come from the hippocampus, and these effects also illustrate the difficulties in making all hormone effects on nerve cells fit the classical mold of "genomic" and "non-genomic" actions. Electrophysiological studies of hippocampal slices initially called attention to rapid effects of gonadal steroids on electrical excitability,[157] raising the possibility of membrane actions. On the other hand, mapping studies of [³H]estradiol uptake in hippocampus showed a sparse distribution of estrophillic neurons containing intracellular estrogen receptors in what appear to be interneurons in the CA1 region as well as other regions of Ammons horn.[158] There was also an indication that the CA1 region of hippocampus contains some estrogen-inducible progesterone receptors, albeit at much lower levels than in the hypothalamus.[159] Yet none of this data, even taken together, indicated that the hippocampus is a major target for estrogen action, compared to brain regions such as the hypothalamus.

What did point in that direction was a study on seizure susceptibility of the dorsal hippocampus during the estrous cycle, in which a marked decrease in seizure threshold was seen on the afternoon of proestrus, at the peak of estrogen and progesterone levels.[160] A morphological study that demonstrated estrogen-induction of dendritic spines and new synapses in the ventromedial hypothalamus of the female rat[161] led to the subsequent discovery that the density of dendritic spines on CA1 pyramidal neurons was also increased by estradiol.[162] Estrogen effects on dendritic spines were found only in the CA1 region and not in the CA3 region or in dentate gyrus.[162] Moreover, spine density changed cyclically during the estrous cycle of the female rat.[163] There were also parallel changes in synapse density on dendritic spines, strongly supporting the notion of synapse turnover during the natural reproductive cycle of the rat.[164]

The critical involvement of progesterone was indicated by the fact that progesterone administration rapidly potentiated estrogen-induced spine formation[162] but then triggered the down-regulation of spines on CA1 neurons.[165] The down-regulation of dendritic spines occurred slowly when estrogen was withdrawn, but took place within 8 to 12 h when progesterone was administered; moreover, the natural down-regulation of dendritic spines between the proestrus peak and the trough on the day of estrus was blocked by the progesterone antagonist, RU38486.[165]

Antagonism of the progesterone effect by RU38486 is consistent with the involvement of intracellular progestin receptors, and is compatible with the finding, noted above, of estrogen-inducible progestin receptors in the CA1 region of hippocampus.[159] However, we know little about the specificity of the progesterone recognition sites in neuronal membranes, and the intracellular progestin receptors have not been localized by immunocytochemistry. In addition, the only detectable intracellular estrogen receptors appear to be in interneurons, not in the CA1 pyramidal neurons themselves.[158] Moreover, estrogen induction of the mRNA for glutamic acid decarboxylase (GAD), a GABA synthetic enzyme, was detected in interneurons residing within the CA1 pyramidal cell layer,[166] a finding consistent with the known localization of estrogen receptors. Could it be that the estrogen effects on formation of synapses on CA1 pyramidal neurons are indirect and mediated through the GABA interneurons? We shall return to this question below.

However, there is another piece to the puzzle, namely, the fact that estrogen induction of new spine synapses on CA1 pyramidal neurons is blocked by concurrent administration of NMDA receptor antagonists, MK801 or

CGP43487.[167] Scopolamine and NBQX, antagonists of muscarinic and non-NMDA excitatory amino acid receptors respectively, were ineffective at doses that interfere with cognition and block ischemic damage, respectively.[167] Spine synapses are excitatory and it is likely that NMDA receptors occur on them.[168,169] Therefore, it is of particular interest that one of the long-term effects of estradiol is to induce NMDA receptor binding sites in the CA1 region of the hippocampus.[170]

When this was found, it was tempting to regard NMDA receptors as a biochemical marker of spine formation. However, NMDA receptors have been shown to mediate important morphological changes in the brain, such as synapse elimination,[171] retracting of spinules of retinal horizontal cells,[172] neuronal migration,[173] and protection from cell death.[174] It is therefore more relevant to ask whether activation of NMDA receptors themselves could lead to induction of new synapses, in which case estrogen induction of NMDA receptors would then become a primary event leading to synapse formation. NMDA receptors gate calcium ions and this may be an important factor in the extension and retraction of dendritic spines.[175] It was found, for example, that NMDA receptor activation promotes dephosphorylation of MAP2 and alters the interaction of this cytoskeletal protein with actin and tubulin.[176–178] There are, of course, many questions to be answered regarding the necessary and sufficient conditions for NMDA receptor activation to induce synapse formation and whether the cascade of events triggered by NMDA receptor activation involves induction of cytoskeletal proteins or of enzymes that modify their interactions as via phosphorylation and dephosphorylation.

We now return to the possible significance of the presence of estrogen receptors in inhibitory interneurons, and not in CA1 pyramidal neurons in which spine formation takes place. Estradiol has a number of effects on the GABA system in the CA1 region. First, there is the estrogen induction of GAD mRNA in inhibitory interneurons within the CA1 pyramidal cell layer.[166] Second, estrogen treatment induces $GABA_A$ receptor binding in the CA1 region of hippocampus.[179] It is possible that this induction of $GABA_A$ receptors is secondary to alterations in GABA transmission in the inhibitory interneurons. In this connection, there are studies pointing to estrogen effects that excite hippocampal CA1 pyramidal neurons, possibly through disinhibition. Wong and Moss[127] have shown recently that 2 d estradiol treatment prolongs the EPSP and increases the probability of repetitive firing in some CA1 neurons in response to Schaffer collateral (glutamatergic) stimulation. One explanation is that estrogen actions on inhibitory interneurons might, in some manner, disinhibit the pyramidal neurons, allowing for removal of the magnesium blockade and activation of NMDA receptors.[167] This would have to occur by disinhibition of the inhibitory influence of other interneurons on pyramidal cell firing, and the neuroanatomical pathway for such an effect remains to be worked out. Moreover, it is not clear how such a sequence of events would lead to induction of NMDA receptors, or for that matter, to induction of $GABA_A$ receptors or, finally, to the induction of new excitatory synapses on dendritic spines.

C. CORTICOSTERONE AND THE SUPPRESSION
OF REPRODUCTIVE BEHAVIOR

Even more dramatic than the increase in progesterone concentrations during proestrus is the increase in circulating corticosterone in response to stress. Corticosterone levels in rats may increase 20-fold within 30 min of the onset of stress. Not surprisingly, considering that stressful situations may be life-threatening, neurons are equipped to respond rapidly to dramatic changes in corticosteroid concentration. It appears that activation of intracellular corticoid receptors may elicit some changes in neuronal excitability that require protein synthesis within 30 to 60 min,[180] although the gene products involved are unknown. Other neuronal responses to corticosteroids, or corticosteroid-BSA complexes, are more rapid[181–184] and may involve membrane receptors for corticosteroids.[56–58] There are several reports of rapidly emerging behavioral responses to corticosteroids,[185–188] but such behaviors have not been extensively studied in rodents.

It has been easier to quantify corticosterone-sensitive, stereotypic behaviors in an amphibian, *Taricha granulosa,* the roughskin newt, in which short-term stress suppresses male reproductive behavior.[189] Male courtship behavior in this species involves protracted amplectic clasping of the female. As in most vertebrates, sexual behavior is modulated by the interaction of steroids and neuropeptides, so that seasonal increases in androgen secretion enhance the responsiveness of males to the actions of the stimulatory neuropeptide, arginine vasotocin (AVT).[190] The behavioral actions of androgens are likely to be related to the actions of testosterone on vasotocin content and receptor number in hypothalamic and extra-hypothalamic brain regions.[191,192] During the breeding season though, this androgen- and AVT-dependent behavior can be suppressed by short-term stress or within 8 min of intraperitoneal injection of corticosterone.[65]

The traditional model of steroid action cannot easily account for the rapidity of onset of this robust behavioral response to corticosterone, and several lines of research support the conclusion that the response is mediated by corticosterone receptors in neuronal membranes. Radioligand binding studies indicated that there are high-affinity recognition sites for [^{3}H]corticosterone in brain membranes that are highly specific for naturally occurring corticosteroids, having low affinity for aldosterone and the synthetic glucocorticoid dexamethasone.[65,67] Subcellular fractionation studies indicated that the specific binding sites were most enriched in synaptic membrane fractions. In addition, [^{3}H]corticosterone, in the presence of unlabeled dexamethasone, was used in *in vitro* receptor autoradiographic studies to label membrane binding sites. Consistent with the fractionation studies, receptors were localized in the neuropil, rather than over the cell bodies, and in regions such as the preoptic area, that are known to play a role in the regulation of sexual behavior.

The binding sites appear to represent functional receptors — behavioral and electrophysiological studies suggested that corticosterone alters male courtship behavior by acting at these receptors. The behavior tests consisted of 20 min trials conducted 5 min after intraperitoneal (ip) injection of steroid to selectively study the rapid corticosteroid effects on male clasping behavior. Clasping was significantly suppressed with ip injection of as little as 1 µg corticosterone, consistent with a high-affinity, membrane site of action. In dose response studies, the potencies of five steroids to inhibit male courtship behavior were found to be strongly correlated with their potencies as inhibitors of [^{3}H]corticosterone binding to neuronal membranes.[65] Extracellular recordings from single cells in the hindbrain demonstrated that ip injection of corticosterone produced multiple electrophysiological effects in two populations of neurons.[193] Corticosterone inhibited the firing of spontaneously active medullary neurons, or decreased their responsiveness to cloacal stimulation. In these animals, cloacal pressure is a stimulus that evokes the male clasping reflex. The responsiveness of these neurons to sensory stimuli was selective, in that the responses to a number of sensory stimuli other than cloacal pressure was not affected by corticosterone. Corticosterone also altered the action potentials of reticulospinal neurons that were antidromically activated from the spinal cord.[193] Similar to the medullary neurons that could not be backfired from the spinal cord, these reticulospinal neurons are responsive to cloacal stimulation before, but not after, corticosterone injection. In both populations of neurons, electrophysiological responses emerged within 3 to 5 minutes after injection, and were maximal within 20 to 30 minutes. Dexamethasone, which does not bind to corticosterone receptors in membranes,[65] did not produce these effects. Together, these behavioral and electrophysiological studies support the notion that corticosterone receptors in brain membranes are functionally relevant and are present in the neuronal circuitry that mediates courtship behavior. Further, behavioral responses to stress may be mediated, in part, by rapid and selective effects of corticosteroids on neuronal processing of sensory information.

How does the short-latency, inhibitory action of corticosterone interact with the facilitatory action of AVT on the clasping reflex? When AVT is applied to the medulla, it elicits effects that are opposite to corticosterone — it enhances neuronal excitability and the responsiveness of neurons to cloacal stimulation. If corticosterone is injected 30 min prior to AVT application, the excitatory effect of AVT is blocked.[194] Surprisingly, if AVT is applied 10 min prior to corticosterone injection, the suppressive effects of corticosterone are blocked, or even reversed. Therefore, the behavioral response to corticosterone cannot be explained solely by the inhibition of AVT release, and there appears to be a complex interaction of corticosterone and AVT that is context dependent. Since GABA$_A$ receptors also have a modulatory role in the clasping reflex, the GABA system may also converge on these hindbrain neurons and effect the steroid-neuropeptide interaction. The GABA$_A$ receptor antagonist, bicuculline, stimulates male clasping when injected intracerebroventricularly, suggesting a tonic inhibitory effect of GABA on courtship behavior.[195] While corticosterone does not directly modulate GABA$_A$ receptors,[69] it interacts with GABA, inasmuch as injection of bicuculline blocks the behavioral action of corticosterone, as does injection of a GABA synthesis inhibitor.[195] We do not know how GABA, AVT and corticosterone interact on a cellular level, but perhaps regulation of membrane potential by GABA modulates the extent of a corticosterone-induced blockade of voltage-dependent Ca^{2+} channels that in some way antagonizes the actions of AVT. Does corticosterone interact with AVT through cross talk between second messenger systems regulated by corticosterone and AVT receptors, or through the uncoupling of AVT receptors to effector mechanisms?

VII. CONCLUSIONS

Vertebrate neurons possess the machinery for dynamic responses to changes in the steroid environment. Basal levels of gonadal and adrenal steroids serve permissive, tonic, and organizing roles in the brain, whereas phasic fluctuations in steroid concentrations may sensitize the brain to respond to reproductive cues or diurnal cycles of activity. Neurons can also respond rapidly to acute changes in steroid concentration, such as the surge of progesterone preceding the onset of behavioral estrus, or the surge of corticosteroids in response to stress. One generalization about signal transduction is that relatively few chemical signals can produce a diversity of neuronal responses through a multitude of receptor mechanisms, and the brain appears to have multiple, temporally segregated transduction pathways for steroids (Figure 4). This generalization is likely to apply to the actions of aldosterone as well, the focus of this volume. Aldosterone has rapid behavioral[196] and electrophysiological[197,198] actions in addition to late-onset genomic actions. These various actions may be mediated through intracellular mineralocorticoid receptors, specific aldosterone receptors in neuronal membranes,[199] or through modulation of $GABA_A$ receptors.

Steroids may have been among the first chemical signals to appear in eukaryotes and multiple mechanisms for steroid action appear to be ancient as well. For example, steroids support gametogenesis in fungi, and both the cytosol and plasma membranes of a filamentous fungus have been shown to contain binding sites for progestins with demonstrably different physiological roles.[200] Even though researchers may have overlooked these ancient mechanisms because of the excitement generated by the discovery that nuclear steroid receptors are transcription factors, these processes tell us that attempts to understand steroid actions in terms of a strict dichotomy between "genomic" and "nongenomic" mechanisms may be fruitless. Rather the examples covered in this chapter should be teaching us that steroids affect brain function via a complex sequence of interacting events initiated by recognition sites in both neuronal membranes and nuclei.

Unfortunately, the receptors mediating the membrane actions of steroids have not been extensively characterized, in part due to the formidable technical problems involved in dealing with steroid binding to the membranous lipid

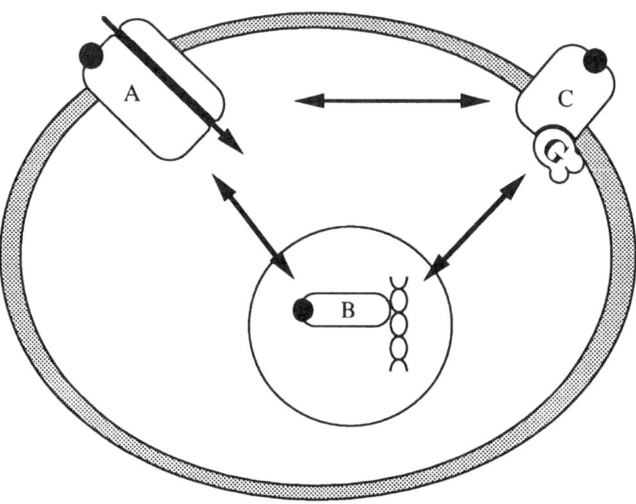

Figure 4. Proposed model for multiple, interacting signal transduction pathways for steroids in the brain. (A) Neuroactive steroids may directly modulate the function of ligand-gated ion channels, such as the $GABA_A$ receptor. Select steroids can hyperpolarize neurons by enhancing GABA-stimulated Cl⁻ conductance. (B) Steroids interact with classical, intracellular receptors to alter the expression of neuron-specific genes. In this manner, steroids modulate neuronal activity by regulating the synthesis or function of neurotransmitters, neuropeptides or a variety of receptors. (C) Steroids may interact with specific receptors in neuronal membranes, thereby altering G protein-regulated processes such as second messenger systems or ion channel conductance. These 3 mechanisms allow neurons to respond to steroids with a range of latencies. It is likely that considerable cross-talk exists between these various genomic and nongenomic mechanisms that allow steroids to exert fine control over neuronal activity. See text for details.

bilayer. However, such characterization is important for many reasons, especially for the development of specific agonists and antagonists that could be used for functional studies and therapeutic purposes. It is our hope that this chapter will serve as an encouragement for further work in this area in spite of the problems and frustrations.

REFERENCES

1. **Evans, R.M.,** The steroid and thyroid hormone receptor superfamily, *Science,* 240, 889, 1988.
2. **Fuller, P.J.,** The steroid receptor superfamily: mechanisms of diversity, *FASEB J.,* 5, 3092, 1991.
3. **Yamamoto, K.,** Steroid receptor regulated transcription of specific genes and gene networks, *Annu. Rev. Gen.,* 19, 209, 1985.
4. **Green, S. and Chambon, P.,** Nuclear receptors enhance our understanding of transcription regulation, *Trends Genet.,* 4, 309, 1988.
5. **Carson-Jurica, M.A., Schrader, W.T., and O'Malley, B.W.,** Steroid receptor family: structure and functions, *Endocr. Rev.,* 11, 201, 1990.
6. **Beato, M.,** Transcriptional control by nuclear receptors, *FASEB J.,* 5, 2044, 1991.
7. **Miner, J.N., Diamond, M.I., and Yamamoto, K.R.,** Joints in the regulatory lattice: composite regulation by steroid receptor-AP1 complexes, *Cell Growth Differ.,* 2, 525, 1991.
8. **Meyer, T.E. and Habener, J.F.,** Cyclic adenosine 3′,5′-monophosphate response element binding protein (CREB) and related transcription-activating deoxyribonucleic acid-binding proteins, *Endocr. Rev.,* 14, 269, 1993.
9. **Pfahl, M.,** Nuclear receptor/AP-1 interaction, *Endocr. Rev.,* 14, 651, 1993.
10. **Power, R.F., Mani, S.K., Codina, J., Conneely, O.M., and O'Malley, B.W.,** Dopaminergic and ligand-independent activation of steroid hormone receptors, *Science,* 254, 1636, 1991.
11. **Power, R.F., Lydon, J.P., Conneely, O.M., and O'Malley, B.W.,** Dopamine activation of an orphan of the steroid receptor superfamily, *Science,* 252, 1546, 1991.
12. **McEwen, B.S., de Kloet, E.R., and Rostene, W.,** Adrenal steroid receptors and actions in the nervous system, *Physiol. Rev.,* 66, 1121, 1986.
13. **McEwen, B.S.,** Non-genomic and genomic effects of steroids on neural activity, *Trends Pharmacol. Sci.,* 12, 141, 1991.
14. **Weeks, J.C., Ed.,** *Steroid Actions on Neurons and Behavior,* Semin. Neurosci., Vol. 3 (6), Saunders/Academic Press, London, 1991.
15. **Pfaff, D.W.,** Features of a hormone-driven defined neural circuit for a mammalian behavior. Principles illustrated, neuroendocrine syllogisms, and multiplicative steroid effects, *Ann. N.Y. Acad. Sci.,* 563, 131, 1989.
16. **McEwen, B.S., Coirini, H., Danielsson, A., Frankfurt, M., Gould, E., Mendelson, S., Schumacher, M., Segarra, A., and Wooley, C.,** Steroid and thyroid hormones modulate a changing brain, *J. Steroid Biochem. Mol. Biol.,* 40, 1, 1991.
17. **Lambert, J.J., Peters, J.A., and Cottrell, G.A.,** Actions of synthetic and endogenous steroids on the GABA$_A$ receptor, *Trends Pharmacol. Sci.,* 8, 224, 1987.
18. **Duval, D., Durant, S., and Homo-Delarche, F.,** Non-genomic effects of steroids, *Biochim. Biophys. Acta,* 737, 409, 1983.
19. **Weiss, D.J. and Gurpide, E.,** Non-genomic effects of estrogens and antiestrogens, *J. Steroid Biochem.,* 31, 671, 1988.
20. **Schumacher, M.,** Rapid membrane effects of steroid hormones: an emerging concept in neuroendocrinology, *Trends Neurosci.,* 13, 359, 1990.
21. **Godeau, J.F., Schorderet-Slatkine, S., Hubert, P., and Baulieu, E.-E.,** Induction of maturation in *Xenopus laevis* oocytes by a steroid linked to a polymer, *Proc. Natl. Acad. Sci. U.S.A.,* 75, 2353, 1978.
22. **Hua, S.Y. and Chen, Y.Z.,** Membrane receptor-mediated electrophysiological effects of glucocorticoid on mammalian neurons, *Endocrinology,* 124, 687, 1989.
23. **Dluzen, D.E. and Ramirez, V.D.,** Progesterone effects upon dopamine release from the corpus striatum of female rats. II. Evidence for a membrane site of action and the role of albumin, *Brain Res.,* 476, 338, 1989.

24. **Petitti, N. and Etgen, A.M.,** Progesterone promotes rapid desensitization of α1-adrenergic receptor augmentation of cAMP formation in rat hypothalamic slices, *Neuroendocrinology,* 55, 1, 1992.

25. **Frye, C.A. and DeBold, J.F.,** P-3-BSA, but not P-11-BSA, implants in the VTA rapidly facilitate receptivity in hamsters after progesterone priming to the VMH, *Behav. Brain Res.,* 53, 167, 1993.

26. **Edwardson, J.A. and Bennett, G.W.,** Modulation of corticotrophin-releasing factor release from hypothalamic synaptosomes, *Nature,* 251, 425, 1974.

27. **Drouva, S.V., Laplante, E., and Kordon, C.,** Progesterone-induced LHRH release *in vitro* is an estrogen — as well as Ca^{2+} — and calmodulin-dependent secretory process, *Neuroendocrinology,* 40, 325, 1985.

28. **Keller-Wood, M.E. and Dallman, M.F.,** Corticosteroid inhibition of ACTH secretion, *Endocr. Rev.,* 5, 1, 1984.

29. **Nabekura, J., Oomura, Y., Minami, T., Mizuno, Y., and Fukuda, A.,** Mechanism of the rapid effect of 17β-estradiol on medial amygdala neurons, *Science,* 233, 226, 1986.

30. **Schumacher, M., Coirini, H., Pfaff, D.W., and McEwen, B.S.,** Behavioral effects of progesterone associated with rapid modulation of oxytocin receptors, *Science,* 250, 691, 1990.

31. **Fink, K.L., Wieben, E.D., Woloschak, G.E., and Spelsberg, T.C.,** Rapid regulation of c-myc protooncogene expression by progesterone in the avian oviduct, *Proc. Natl. Acad. Sci. U.S.A.,* 85, 1796, 1988.

32. **Bitran, D., Purdy, R.H., and Kellogg, C.K.,** Anxiolytic effect of progesterone is associated with increases in cortical allopregnanolone and $GABA_A$ receptor function, *Pharmacol. Biochem. Behav.,* 45, 423, 1993.

33. **Rupprecht, R., Reul, J.M.H.M., Trapp, T., van Steensel, B., Wetzel, C., Damm, K., Zieglgansberger, W., and Holsboer, F.,** Progesterone receptor-mediated effects of neuroactive steroids, *Neuron,* 11, 523, 1993.

34. **Arnold, A.P. and Gorski, R.A.,** Gonadal steroid induction of structural sex differences in the central nervous system, *Annu. Rev. Neurosci.,* 7, 413, 1984.

35. **Schlinger, B.A. and Arnold, A.P.,** Brain is the major site of estrogen synthesis in a male songbird, *Proc. Natl. Acad. Sci. U.S.A.,* 88, 4191, 1991.

36. **Schlinger, B.A. and Callard, G.V.,** Localization of aromatase in synaptosomal and microsomal subfractions of quail *(Coturnix coturnix japonica)* brain, *Neuroendocrinology,* 49, 434, 1989.

37. **Selmanoff, M.K., Bradkin, L.D., Weiner, R.I., and Siiteri, P.K.,** Aromatization and 5α-reduction of androgens in discrete hypothalamic and limbic regions of the male and female rat, *Endocrinology,* 101, 841, 1977.

38. **Krieger, N.R. and Scott, R.G.,** 3α-Hydroxysteroid oxidoreductase in rat brain, *J. Neurochem.,* 42, 887, 1984.

39. **Karavolas, H.J. and Hodges, D.R.,** Neuroendocrine metabolism of progesterone and related progestins, in *Steroids and Neuronal Activity,* Ciba Found. Symp. 153, Chadwick, D. and Widdows, K., Eds., John Wiley and Sons, New York, 1990, 22.

40. **Hu, Z.Y., Bourreau, E., Jung-Testas, I., Robel, P., and Baulieu, E.-E.,** Neurosteroids: oligodendrocyte mitochondria convert cholesterol to pregnenolone, *Proc. Natl. Acad. Sci. U.S.A.,* 84, 8215, 1987.

41. **Baulieu, E.-E.,** Neurosteroids: a new function in the brain, *Biol. Cell,* 71, 3, 1991.

42. **Mellon, S.H. and Deschepper, C.F.,** Neurosteroid biosynthesis: genes for adrenal steroidogenic enzymes are expressed in the brain, *Brain Res.,* 629, 283, 1993.

43. **Corpechot, C., Synguelakis, M., Talha, S., Axelson, M., Sjovall, J., Vihko, R., Baulieu, E.-E., and Robel, P.,** Pregnenolone and its sulfate ester in the rat brain, *Brain Res.,* 270, 119, 1983.

44. **Korneyev, A., Pan, B.S., Polo, A., Romero, E., Guidotti, A., and Costa, E.,** Stimulation of brain pregnenolone synthesis by mitochondrial diazepam binding inhibitor receptor ligands *in vivo, J. Neurochem.,* 61, 1515, 1993.

45. **Tabel, T. and Heinrichs, W.L.,** Metabolism of progesterone by the brain and pituitary gland of subhuman primates, *Neuroendocrinology,* 15, 281, 1974.

46. **Korneyev, A., Guidotti, A., and Costa, E.,** Regional and interspecies differences in brain progesterone metabolism, *J. Neurochem.,* 61, 2041, 1993.

47. **Lan, N.C., Bolger, M.B., and Gee, K.W.,** Identification and characterization of a pregnane steroid recognition site that is functionally coupled to an expressed $GABA_A$ receptor, *Neurochem. Res.,* 16, 347, 1991.

48. **Harrison, N.L., Majewska, M.D., Harrington, J.W., and Barker, J.L.,** Structure-activity relationships for steroid interaction with the γ-aminobutyric $acid_A$ receptor complex, *J. Pharmacol. Exp. Ther.,* 241, 346, 1987.

49. **Purdy, R.H., Morrow, A.L., Blinn, J.R., and Paul, S.M.,** Synthesis, metabolism, and pharmacological activity of 3α-hydroxy steroids which potentiate GABA-receptor-mediated chloride ion uptake in rat cerebral cortical synaptoneurosomes, *J. Med. Chem.,* 33, 1572, 1990.

50. **Allera, A. and Wildt, L.,** Glucocorticoid-recognizing and -effector sites in rat liver plasma membrane. Kinetics of corticosterone uptake by isolated membrane vesicles. I. Binding and transport, *J. Steroid Biochem. Mol. Biol.,* 42, 737, 1992.

51. **Suyemitsu, T. and Terayama, H.,** Specific binding sites for natural glucocorticoids in plasma membranes of rat liver, *Endocrinology,* 96, 1499, 1975.

52. **Pietras, R.J. and Szego, C.M.,** Specific binding sites for oestrogen at the outer surfaces of isolated endometiral cells, *Nature,* 265, 69, 1977.

53. **Koch, B., Lutz-Bucher, B., Briaud, B., and Mialhe, C.,** Specific interaction of corticosteroids with binding sites in the plasma membranes of the rat anterior pituitary gland, *J. Endocrinol.,* 79, 215, 1978.

54. **Harrison, R.W., Balasubramanian, K., Yeakley, J., Fant, M., Svec, F., and Fairfield, S.,** Heterogeneity of AtT-20 cell glucocorticoid binding sites: evidence for a membrane receptor, in *Steroid Hormone Receptor Systems,* Leavitt, W.W. and Clark, J.H., Eds., Plenum Publishing, New York, 1979, 423.

55. **Orchinik, M. and McEwen, B.S.,** Novel and classical actions of neuroactive steroids, *Neurotransmissions,* 9, 1, 1993.

56. **Towle, A.C. and Sze, P.Y.,** Steroid binding to synaptic plasma membrane: differential binding of glucocorticoids and gonadal steroids, *J. Steroid Biochem.,* 18, 135, 1983.

57. **Chen, Y.-Z., Fu, H., and Guo, Z.,** Membrane receptors for glucocorticoids in mammalian neurons, in *Receptors: Model Systems and Specific Receptors,* Meth. Neurosci., Vol. 11, Conn, P.M., Ed., Academic Press, New York, 1993, chap. 2.

58. **Sze, P.Y. and Towle, A.C.,** Developmental profile of glucocorticoid binding to synaptic plasma membrane from rat brain, *Int. J. Dev. Neurosci.,* 11, 339, 1993.

59. **Dallman, M.F., Akana, S.K., Cascio, C.S., Darlington, D.N., Jacobson, L., and Levin, N.,** Regulation of ACTH secretion: variations on a theme of B, *Recent Prog. Horm. Res.,* 43, 113, 1987.

60. **De Kloet, E.R., Reul, J.M.H.M., de Ronde, F.S.W., and Ratka, A.,** Function and plasticity of brain corticosteroid receptor systems: action of neuropeptides, *J. Steroid Biochem.,* 25, 723, 1986.

61. **Funder, J.W.,** Glucocorticoid receptors, *J. Ster. Biochem. Mol. Biol.,* 43, 389, 1992.

62. **Liposits, Z. and Bohn, M.C.,** Association of glucocorticoid receptor immunoreactivity with cell membrane and transport vesicles in hippocampal and hypothalamic neurons of the rat, *J. Neurosci. Res.,* 35, 14, 1993.

63. **Blaustein, J.D., Lehman, M.N., Turcotte, J.C., and Greene, G.,** Estrogen receptors in dendrites and axon terminals in the guinea pig hypothalamus, *Endocrinology,* 131, 281, 1992.

64. **Skipper, J.K., Young, L.J., Bergerson, J.M., Tetzlaff, M.T., Osborn, C.T., and Crews, D.,** Identification of an isoform of the estrogen receptor messenger RNA lacking exon four and present in the brain, *Proc. Natl. Acad. Sci. U.S.A.,* 90, 7172, 1993.

65. **Orchinik, M., Murray, T.F., and Moore, F.L.,** A corticosteroid receptor in neuronal membranes, *Science,* 252, 1848, 1991.

66. **Orchinik, M., Murray, T.F., Franklin, P.H., and Moore, F.L.,** Guanyl nucleotides modulate binding to steroid receptors in neuronal membranes, *Proc. Natl. Acad. Sci. U.S.A.,* 89, 3830, 1992.

67. **Orchinik, M., Murray, T.F., and Moore, F.L.,** Corticosteroid receptor in neuronal membranes associated with rapid suppression of sexual behavior, in *Neurosteroids and Brain Function,* Fidia Research Foundation Symposia Series, Vol. 8, Costa, E. and Paul, S.M., Eds., Thieme Medical Publishers, New York, 1991, 125.

68. **Orchinik, M. and Murray, T.F.,** Steroid Binding to Membrane Receptors, in *Neurobiology of Steroids, Methods in Neuroscience,* Vol. 22, Academic Press, 1994, in press.

69. **Orchinik, M., Murray, T.F., and Moore, F.L.,** Steroid modulation of GABA$_A$ receptors in an amphibian brain, *Brain Res.,* in press 1994.

70. **Majewska, M.D., Demirgoren, S., and London, E.D.,** Binding of pregnenolone sulfate to rat brain membranes suggests multiple sites of steroid action at the GABA$_A$ receptor, *Eur. J. Pharmacol. Mol. Pharm. Sec.,* 189, 307, 1990.

71. **Majewska, M.D., Demirgoren, S., Spivak, C.E., and London, E.D.,** The neurosteroid dehydroepiandrosterone sulfate is an allosteric antagonist of the GABA$_A$ receptor, *Brain Res.,* 526, 143, 1990.

72. **Demirgoren, S., Majewska, M.D., Spivak, C.E., and London, E.D.,** Receptor binding and electrophysiological effects of dehydroepiandrosterone sulfate, an antagonist of the GABA$_A$ receptor, *Neuroscience,* 45, 127, 1991.

73. **Ke, F.-C. and Ramirez, V.D.,** Binding of progesterone to nerve cell membranes of rat brain using progesterone conjugated to ^{125}I-bovine serum albumin as ligand, *J. Neurochem.,* 54, 467, 1990.

74. **Tischkau, S.A. and Ramirez, V.D.,** A specific membrane binding protein for progesterone in rat brain: sex differences and induction by estrogen, *Proc. Natl. Acad. Sci. U.S.A.,* 90, 1285, 1993.

75. **Ke, F.-C. and Ramirez, V.D.,** Membrane mechanism mediates progesterone stimulatory effect on LHRH release from superfused rat hypothalami *in vitro, Neuroendocrinology,* 45, 514, 1987.

76. **Orchinik, M., Witt, D.M., and McEWen, B.S.,** unpublished data, 1993.

77. **Rosner, W.,** The functions of corticosteroid-binding globulin and sex hormone binding globulin: recent advances, *Endocr. Rev.,* 11, 80, 1990.

78. **Strel'chyonok, O.A. and Avvakumov, G.V.,** Interaction of human CBG with cell membranes, *J. Steroid Biochem.,* 40, 795, 1991.

79. **Maitra, U.S., Khan, M.S., and Rosner, W.,** Corticosteroid-binding globulin receptor of the rat hepatic membrane: solubilization, partial characterization, and the effect of steroids on binding, *Endocrinology,* 133, 1817, 1993.

80. **Schofield, P.R., Darlison, M.G., Fujita, N., Burt, D.R., Stephenson, F.A., Rodriguez, H., Rhee, L.M., Ramachandran, J., Reale, V., Glencorse, T.A., Seeburg, P.H., and Barnard, E.A.,** Sequence and functional expression of the GABA$_A$ receptor shows a ligand-gated receptor super-family, *Nature,* 328, 221, 1987.

81. **Bellelli, D., Lan, N.C., and Gee, K.W.,** Anticonvulsant steroids and the GABA/benzodiazepine receptor-chloride ionophore complex, *Neurosci. Biobehav. Rev.,* 14, 315, 1990.

82. **Paul, S.M. and Purdy, R.H.,** Neuroactive steroids, *FASEB J.,* 6, 2311, 1992.

83. **Majewska, M.D.,** Neurosteroids: endogenous bimodal modulators of the GABA$_A$ receptor. Mechanisms of action and physiological significance, *Prog. Neurobiol.,* 38, 379, 1992.

84. **Harrison, N.L., Majewska, M.D., Meyers, D.E.R., and Barker, J.L.,** Rapid actions of steroids on CSN neurons, in *Neural Control of Reproductive Function,* Lakoski, J.M., Perez-Polo, J.R., and Rassin, D.K., Eds., Alan R. Liss, New York, 1989, 137.

85. **Myslobodsky, M.S.,** Pro- and anticonvulsant effects of stress: the role of neuroactive steroids, *Neurosci. Biobehav. Rev.,* 17, 129, 1993.

86. **Puia, G., Santi, M., Vicini, S., Pritchett, D.B., Purdy, R.H., Paul, S.M., Seeburg, P.H., and Costa, E.,** Neurosteroids act on recombinant human GABA$_A$ receptors, *Neuron,* 4, 759, 1990.

87. **Lan, N.C., Chen, J.-S., Belelli, D., Pritchett, D.B., Seeburg, P.H., and Gee, K.W.,** A steroid recognition site is functionally coupled to an expressed GABA$_A$-benzodiazepine receptor, *Mol. Pharmacol.,* 188, 403, 1990.

88. **Shingai, R., Sutherland, M.L., and Barnard, E.A.,** Effects of subunit types of the cloned GABA$_A$ receptor on the response to a neurosteroid, *Eur. J. Pharmacol. Mol. Pharm.,* 206, 77, 1991.

89. **Hill-Venning, C., Lambert, J.J., Peters, J.A., and Hales, T.G.,** The actions of neurosteroids on inhibitory amino acid receptors, in *Neurosteroids and Brain Function,* Fidia Res. Found. Symp. Ser., Vol. 8, Costa, E. and Paul, S.M., Eds., Thieme Medical Publishers, New York, 1991, 77.

90. **Lan, N.C., Gee, K.W., Bolger, M.B., and Chen, J.S.,** Differential responses of expressed recombinant human γ-aminobutyric acid$_A$ receptors to neurosteroids, *J. Neurochem.,* 57, 1818, 1991.

91. **Woodward, R.M., Polenzani, L., and Miledi, R.,** Effects of steroids on γ-aminobutyric acid receptors expressed in *Xenopus* oocytes by poly(A)$^+$ RNA from mammalian brain and retina, *Mol. Pharmacol.,* 41, 89, 1992.

92. **Lambert, J.J., Peters, J.A., Sturgess, N.C., and Hales, T.G.,** Steroid modulation of the GABA$_A$ receptor complex: electrophysiological studies, in *Steroids and Neuronal Activity,* Ciba Found. Symp. 153, Chadwick, D. and Widdows, K., Eds., John Wiley and Sons, New York, 1990, 56.

93. **Giusti, L., Belfiore, M.S., Martini, C., and Lucacchini, A.,** 3α-Hydroxy-5α-pregnan-20-one modulation of solubilized GABA/benzodiazepine receptor complex, *J. Ster. Biochem. Mol. Biol.,* 45, 309, 1993.

94. **Burt, D.R. and Kamatchi, G.L.,** GABA$_A$ receptor subtypes: from pharmacology to molecular biology, *FASEB J.,* 5, 2916, 1991.

95. **Olsen, R.W. and Tobin, A.J.,** Molecular biology of GABA$_A$ receptors, *FASEB J.,* 4, 1469, 1990.

96. **Seeburg, P.H., Wisden, W., Verdoorn, T.A., Pritchett, D.B., Werner, P., Herb, A., Luddens, H., Sprengel, R., and Sakmann, B.,** The GABA$_A$ receptor family: molecular and functional diversity, *Cold Spring Harbor Symp. on Quant. Biol.,* 55, 29, 1990.

97. **Sieghart, W.,** GABA$_A$ receptors: ligand-gated Cl$^-$ ion channels modulated by multiple drug-binding sites, *Trends Pharmacol. Sci.,* 13, 446, 1992.

98. **Schmitz, E., Reichelt, R., Walkowiak, W., Richards, J.G., and Hebebrand, J.,** A comparative phylogenetic study of the distribution of cerebellar GABA$_A$/benzodiazepine receptors using radioligands and monoclonal antibodies, *Brain Res.,* 473, 314, 1988.

99. **Zaman, S.H., Shingai, R., Harvey, R.J., Darlison, M.G., and Barnard, E.A.,** Effects of subunit types of the recombinant GABA$_A$ receptor on the response to a neurosteroid, *Eur. J. Pharmacol. Mol. Pharm. Sec.,* 225, 321, 1992.

100. **Gee, K.W., Bolger, M.B., Brinton, R.E., Coirini, H., and McEwen, B.S.,** Steroid modulation of the chloride ionophore in rat brain: structure-activity requirements, regional dependence and mechanism of action, *J. Pharmacol. Exp. Ther.,* 246, 803, 1988.

101. **Gee, K.W. and Lan, N.C.,** γ-Aminobutyric acid$_A$ receptor complexes in rat frontal cortex and spinal cord show differential responses to steroid modulation, *Mol. Pharmacol.,* 40, 995, 1991.

102. **Sapp, D.W., Witte, U., Turner, D.M., Longoni, B., Kokka, N., and Olsen, R.W.,** Regional variation in steroid anesthetic modulation of [^{35}S]TBPS binding to γ-aminobutyric acid$_A$ receptors in rat brain, *J. Pharmacol. Exp. Ther.,* 262, 801, 1992.

103. **Jussofie, A.,** Brain region-specific effects of neuroactive steroids on the affinity and density of the GABA-binding site, *Biol. Chem. Hoppe-Seyler,* 374, 265, 1993.

104. **Canonaco, M., Tavolaro, R., and Maggi, A.,** Steroid hormones and receptors of the GABA$_A$ supramolecular complex. II. Progesterone and estrogen inhibitory effects on the chloride ion channel receptor in different forebrain areas of the female rat, *Neuroendocrinology,* 57, 974, 1993.

105. **Ennis, C. and Minchin, M.C.W.,** Modulation of the GABA$_A$-like autoreceptor by barbiturates but not by steroids, *Neuropharmacology,* 32, 355, 1993.

106. **Prince, R.J. and Simmonds, M.A.,** Steroid modulation of the strychnine-sensitive glycine receptor, *Neuropharmacology,* 31, 201, 1992.

107. **Wu, F.-S., Gibbs, T.T., and Farb, D.H.,** Inverse modulation of gamma-aminobutyric acid- and glycine-induced currents by progesterone, *Mol. Pharmacol.,* 37, 597, 1990.

108. **Wu, F.-S., Gibbs, T.T., and Farb, D.H.,** Pregnenolone sulfate: a positive allosteric modulator at the N-methyl-D-aspartate receptor, *Mol. Pharmacol.,* 40, 333, 1991.

109. **Irwin, R.P., Maragakis, N.J., Rogawski, M.A., Purdy, R.H., Farb, D.H., and Paul, S.M.,** Pregnenolone sulfate augments NMDA receptor mediated increases in intracellular Ca^{2+} in cultured rat hippocampal neurons, *Neurosci. Lett.,* 141, 30, 1992.

110. **Meyer, J.H. and Gruol, D.L.,** Dehydroepiandrosterone sulfate alters synaptic potentials in area CA1 of the hippocampal slice, *Brain Res.,* 633, 253, 1994.

111. **Bertrand, D., Valera, S., Bertrand, S., Ballivet, M., and Rungger, D.,** Steroids inhibit nicotinic acetylcholine receptors, *NeuroReport,* 2, 277, 1991.

112. **Valera, S., Ballivet, M., and Bertrand, D.,** Progesterone modulates a neuronal nicotinic acetylcholine receptor, *Proc. Natl. Acad. Sci. U.S.A.,* 89, 9949, 1992.

113. **Mermelstein, P.G., Becker, J.B., Surmeier, D.J.,** 17β-Estradiol inhibits ω-conotoxin sensitive (N-type) calcium channels in rat striatal neurons, *Soc. Neurosci. Abstr.,* 19, 1527, 1993.

114. **ffrench-Mullen, J.M.H. and Spence, K.T.,** Neurosteroids block Ca^{2+} channel current in freshly isolated hippocampal CA1 neurons, *Eur. J. Pharmacol.,* 202, 269, 1991.

115. **ffrench-Mullen, J.M.H., Danks, P., and Spence, K.T.,** Neurosteroids modulate calcium currents in hippocampal CA1 neurons via a pertussis toxin-sensitive G protein-coupled mechanism, *J. Neurosci.,* 14, 1963, 1994.

116. **Hille, B.,** G protein coupling mechanisms and nervous system signaling, *Neuron,* 9, 187, 1992.

117. **Swartz, K.J.,** Modulation of Ca^{2+} channels by protein kinase C in rat central and peripheral neurons: disruption of G protein-mediated inhibition, *Neuron,* 11, 305, 1993.

118. **Becker, J.B.,** Direct effect of 17β-estradiol on striatum: sex differences in dopamine release, *Synapse,* 5, 157, 1990.

119. **Emons, G., Frevert, E.U., Ortmann, O., Fingscheidt, U., Sturm, R., Kiesel, L., and Knuppen, R.,** Studies on the subcellular mechanisms mediating the negative estradiol effect on GnRH-induced LH-release by rat pituitary cells in culture, *Acta Endocrinol.,* 121, 350, 1989.

120. **Nakano, Y., Suda, T., Sumitomo, T., Tozawa, F., and Demura, H.,** Effects of sex steroids on β-endorphin release from rat hypothalamus *in vitro, Brain Res.,* 553, 1, 1991.

121. **Bennett, G.W., Edwardson, J.A., Holland, D., Jeffcoate, S.L., and White, N.,** Release of immunoreactive luteinising hormone-releasing hormone and thyrotropin-releasing hormone rom hypothalamic synaptosomes, *Nature,* 257, 323, 1975.

122. **Gilad, G.M., Rabey, J.M., and Gilad, V.H.,** Presynaptic effects of glucocorticoids on dopaminergic and cholinergic synaptosomes. Implications for rapid endocrine-neural interactions in stress, *Life Sci.,* 40, 2401, 1987.

123. **Meiri, H.,** Is synaptic transmission modulated by progesterone?, *Brain Res.,* 385, 193, 1986.

124. **Borski, R.J., Helms, L.M.H., Richman, N.H., III, and Grau, E.G.,** Cortisol rapidly reduces prolactin release and cAMP and ^{45}Ca^{2+} accumulation in the cichlid fish pituitary in vitro, *Proc. Natl. Acad. Sci. U.S.A.,* 88, 2758, 1991.

125. **Minami, T., Oomura, Y., Nabekura, J., and Fukuda, A.,** 17β-Estradiol depolarization of hypothalamic neurons is mediated by cyclic AMP, *Brain Res.,* 519, 301, 1990.

126. **Smith, S.S.,** The effects of estrogen and progesterone on GABA and glutamate responses at extrahypothalamic sites, in *Neurosteroids and Brain Function*, Fidia Res. Found. Symp. Ser., Vol. 8, Costa, E. and Paul, S.M., Eds., Thieme Medical Publishers, New York, 1991, 87.

127. **Wong, M. and Moss, R.L.,** Long-term and short-term electrophysiological effects of estrogen on the synaptic properties of hippocampal CA1 neurons, *J. Neurosci.,* 12, 3217, 1992.

128. **Lagrange, A.H., Ronnekleiv, O.K., Resko, J.A., and Kelly, M.J.,** Estrogen's rapid modulation of μ-opioid potency in guinea pig hypothalamic arcuate neurons, *Soc. Neurosci. Abstr.,* 19, 1160, 1993.

129. **Kelly, M.J., Loose, M.D., and Ronnekleiv, O.K.,** Estrogen suppresses μ-opioid- and $GABA_B$-mediated hyperpolarization of hypothalamic arcuate neurons, *J. Neurosci.,* 12, 2745, 1992.

130. **Finidori-Lepicard, J., Schorderet-Slatkine, S., Hanoune, J., and Baulieu, E.E.,** Progesterone inhibits membrane-bound adenylate cyclase in *Xenopus laevis* oocytes, *Nature,* 292, 255, 1981.

131. **Sadler, S.E. and Maller, J.L.,** Progesterone inhibits adenylate cyclase in *Xenopus* oocytes, *J. Biol. Chem.,* 256, 6368, 1981.

132. **Smith, L.D.,** The induction of oocyte maturation: transmembrane signaling events and regulation of the cell cycle, *Development,* 107, 685, 1989.

133. **Cork, R.J., Taylor, M., Varnold, R.L., Smith, L.D., and Robinson, K.R.,** Microinjected GTP-γ-S inhibits progesterone-induced maturation of *Xenopus* oocytes, *Dev. Biol.,* 141, 447, 1990.

134. **Chien, E.J., Morrill, G.A., and Kostellow, A.B.,** Progesterone-induced second messengers at the onset of meiotic maturation in the amphibian oocyte: interrelationships between phospholipid N-methylation, calcium and diacylglycerol release, and inositol phospholipid turnover, *Mol. Cell Endocrinol.,* 81, 53, 1991.

135. **Thomas, P. and Meizel, S.,** Phosphatidylinositol 4,5-bisphosphate hydrolysis in human sperm stimulated with follicular fluid or progesterone is dependent upon Ca^{2+} influx, *Biochem. J.,* 264, 539, 1989.

136. **Blackmore, P.F., Neulen, J., Lattanzio, F., and Beebe, S.J.,** Cell surface-binding sites for progesterone mediated calcium uptake in human sperm, *J. Biol. Chem.,* 266, 18655, 1991.

137. **Ravindra, R. and Aronstam, R.S.,** Progesterone, testosterone and estradiol-17β inhibit gonadotropin-releasing hormone stimulation of G protein GTPase activity in plasma membranes from rat anterior pituitary lobe, *Acta Endocrinol.,* 126, 345, 1992.

138. **Bergamini, C.M., Pansini, F., Bettocchi, S., Segala, V., Dallocchio, F., Bagni, B., and Mollica, G.,** Hormonal sensitivity of adenylate cyclase from human endometrium: modulation by estradiol, *J. Steroid Biochem.,* 22, 299, 1985.

139. **Huang, R.-R.C., Dehaven R.N., Cheung, A.H., Diehl, R.E., Dixon, R.A.F., and Strader, C.D.,** Identification of allosteric antagonists of receptor-guanine nucleotide-binding protein interactions, *Mol. Pharmacol.,* 37, 304, 1990.

140. **Mukai, H., Munekata, E., and Higashijima, T.,** G protein antagonists. A novel hydrophobic peptide competes with receptor for G protein binding, *J. Biol. Chem.,* 267, 16237, 1992.

141. **Aridor, M., Rajmilevich, G., Beaven, M.A., and Sagi-Eisenberg, R.,** Activation of exocytosis by the heterotrimeric G protein G_{i3}, *Science,* 262, 1569, 1993.

142. **Sokolovsky, M., Egozi, Y., and Avissar, S.,** Molecular regulation of receptors: interaction of β-estradiol and progesterone with the muscarinic system, *Proc. Natl. Acad. Sci. U.S.A.,* 78, 5554, 1981.

143. **Levesque, D. and Di Paolo, T.,** Rapid conversion of high into low striatal D_2-dopamine receptor agonist binding states after an acute physiological dose of 17β-estradiol, *Neurosci. Lett.,* 88, 113, 1988.

144. **Zhang, G., Murray, T.F., and Kelly, M.J.,** Effect of estrogen on μ-opioid receptor G-protein coupling in guinea pig mediobasal hypothalamus, *Soc. Neurosci. Abstr.,* 19, 1154, 1993.

145. **Pfaff, D.W. and Schwartz-Giblin, S.,** Cellular mechanisms of female reproductive behavior, in *The Physiology of Reproduction,* Knobil, E. and J. Neill et. al., Eds., Raven Press, New York, 1988, 1487.

146. **Etgen, A.M., Ungar, S., and Pettiti, N.,** Estradiol and progesterone modulation of norepinephrine neurotransmission: implications for the regulation of female reproductive behavior, *J. Neuroendocrinol.,* 4, 255, 1992.

147. **Havens, M.D. and Rose, J.D.,** Estrogen-dependent and estrogen-independent effects of progesterone on the electrophysiological excitability of dorsal midbrain neurons in golden hamsters, *Neuroendocrinology,* 48, 120, 1988.

148. **Rose, J.D.,** Changes in hypothalamic neuronal function related to hormonal induction of lordosis in behaving hamsters, *Physiol. Behav.,* 47, 1201, 1990.

149. **Rose, J.D.,** Forebrain influences on brainstem and spinal mechanisms of copulatory behavior: a current perspective on Frank Beach's contribution, *Neurosci. Biobehav. Rev.,* 14, 207, 1990.

150. **Frye, C.A., Mermelstein, P.G., and DeBold, J.F.,** Evidence for a non-genomic action of progestins on sexual receptivity in hamster ventral tegmental area but not hypothalamus, *Brain Res.,* 578, 87, 1992.

151. **Frye, C.A. and Leadbetter, E.A.,** 5α-Reduced progesterone metabolites are essential in hamster VTA for sexual receptivity, *Life Sci.,* 1994, in press.

152. **Frye, C.A. and DeBold, J.F.,** 3α-OH-DHP and 5α-THDOC implants to the VTA facilitate sexual receptivity in hamsters after progesterone priming to the VMH, *Brain Res.,* 612, 130, 1992.

153. **Schumacher, M., Coirini, H., and McEwen, B.S.,** Regulation of high-affinity $GABA_A$ receptors in specific brain regions by ovarian hormones, *Neuroendocrinology,* 50, 315, 1989.

154. **O'Connor, L.H., Nock, B., and McEwen, B.S.,** Regional specificity of gamma-aminobutyric acid receptor regulation by estradiol, *Neuroendocrinology,* 47, 473, 1988.

155. **Juptner, M., Jussofie, A., and Hiemke, C.,** Effects of ovariectomy and steroid replacement on $GABA_A$ receptor binding in female rat brain, *J. Ster. Biochem. Molec. Biol.,* 38, 141, 1991.

156. **Kelly, M.J., Moss, R.L., and Dudley, C.A.,** Differential sensitivity of preoptic-septal neurons to microelectrophoresed estrogen during the estrous cycle, *Brain Res.,* 114, 152, 1976.

157. **Teyler, T.J., Vardaris, R.M., Lewis, D., and Rawitch, A.B.,** Gonadal steroids: effects on excitability of hippocampal pyramidal cells, *Science,* 209, 1017, 1980.

158. **Loy, R., Gerlach, J.L., and McEwen, B.S.,** Autoradiographic localization of estradiol-binding neurons in the hippocampal formation and the entorhinal cortex, *Dev. Brain Res.,*39, 245, 1988.

159. **Parsons, B., Rainbow, T.C., MacLusky, N., and McEwen, B.S.,** Progestin receptor levels in rat hypothalamic and limbic nuclei, *J. Neurosci.,* 2, 1446, 1982.

160. **Terasawa, E. and Timeras, P.,** Electrical activity during the estrous cycle of the rat:cyclic changes in limbic structures, *Endocrinology,* 83, 207, 1968.

161. **Frankfurt, M., Gould, E., Woolley, C., and McEwen, B.S.,** Gonadal steroids modify dendritic spine density in ventromedial hypothalamic neurons:a golgi study in the adult rat, *Neuroendocrinology,* 51, 530, 1990.

162. **Gould, E., Woolley, C., Frankfurt, M., and McEwen, B.S.,** Gonadal steroids regulate dendritic spine density in hippocampal pyramidal cells in adulthood, *J. Neurosci.,* 10, 1286, 1990.

163. **Woolley, C., Gould, E., Frankfurt, M., and McEwen, B.S.,** Naturally occurring fluctuation in dendritic spine density on adult hippocampal pyramidal neurons, *J. Neurosci.,* 10, 4035, 1990.

164. **Woolley, C. and McEwen, B.S.,** Estradiol mediates naturally-occurring fluctuation in adult hippocampal synapse density, *J. Neurosci.,* 12, 2549, 1992.

165. **Woolley, C. and McEwen, B.S.,** Roles of estradiol and progesterone in regulation of hippocampal dendritic spine density during the estrous cycle in the rat, *J. Comp. Neurol.,* 336, 293, 1993.

166. **Weiland, N.G.,** Glutamic acid decarboxylase messenger ribonucleic acid is regulated by estradiol and progesterone in the hippocampus, *Endocrinology,* 131, 2697, 1992.

167. **Woolley, C. and McEwen, B.S.,** Estradiol regulates hippocampal dendritic spine density via an NMDA receptor dependent mechanism, *J. Neurosci.,* in press 1994.

168. **Horner, C.,** Plasticity of the dendritic spine, *Prog. Neurobiol.,* 41, 281, 1993.

169. **Muller, W. and Connor, J.,** Dendritic spines as individual neuronal compartments for synaptic Ca^{2+} responses, *Nature,* 354, 73, 1991.

170. **Weiland, N.G.,** Estradiol selectively regulates agonist binding sites on the N-methyl-D-aspartate receptor complex in the CA1 region of the hippocampus, *Endocrinology,* 131, 662, 1992.

171. **Rabacchi, S., Bailly, Y., Delhaye-Bouchaud, N., and Mariani, J.,** Involvement of the N-methyl-D-aspartate (NMDA) receptor in synapse elimination during cerebellar development, *Science,* 256, 1823, 1992.

172. **Weiler, R. and Schultz, K.,** Ionotropic non-N-methyl-D-aspartate agonists induce retraction of dendritic spinules from retinal horizontal cells, *Proc. Natl. Acad. Sci. U.S.A.,* 90, 6533, 1993.

173. **Komuro, H. and Rakic, P.,** Modulation of neuronal migration by NMDA receptors, *Science,* 260, 95, 1993.

174. **Gould, E., Cameron, H.A., and McEwen, B.S.,** Blockade of NMDA receptors increases cell death and birth in the developing dentate gyrus, *J. Comp. Neurol.,* 1993, in press.

175. **Benke, T., Jones, O., Collingridge, G., and Angelides, K.,** N-methyl-D-aspartate receptors are clustered and immobilized on dendrites of living cortical neurons, *Proc. Natl. Acad. Sci. U.S.A.,* 90, 7819, 1993.

176. **Halpain, S. and Greengard, P.,** Activation of NMDA receptors induces rapid dephosphorylation of cytoskeletal protein MAP2, *Neuron,* 5, 237, 1990.

177. **Murthy, A.S. and Flavin, M.,** Microtubule assembly using the microtubule associated protein MAP2 prepared in defined states of phosphorylation with protein kinase and phosphatase, *Eur. J. Biochem.,* 137, 37, 1983.

178. **Seldon, S. and Pollard, T.,** Phosphorylation of microtubule associated proteins regulates their interaction with actin filaments, *J. Biol. Chem.,* 258, 7064, 1983.

179. **Schumacher, M., Coirini, H., and McEwen, B.S.,** Regulation of high-affinity $GABA_A$ receptors in the dorsal hippocampus by estradiol and progesterone, *Brain Res.,* 484, 178, 1989.

180. **Karst, H. and Joels, M.,** The induction of corticosteroid actions on membrane properties of hippocampal CA1 neurons requires protein synthesis, *Neurosci. Lett.,* 130, 27, 1991.

181. **Kelly, M.J., Moss, R.L., and Dudley, C.A.,** The effects of microelectrophoretically applied estrogen, cortisol, and acetylcholine on medial preoptic septal unit activity throughout the estrous cycle of the female rat, *Exp. Brain Res.,* 30, 53, 1977.

182. **Saphier, D. and Feldman, S.,** Iontophoretic application of glucocorticoids inhibits identified neurones in the rat paraventricular nucleus, *Brain Res.,* 453, 183, 1988.

183. **Chen, Y.Z., Hua, S.Y., Wang, C.A., Wu, L.G., Gu, Q., and Xing, B.R.,** An electrophysiological study on the membrane receptor-mediated action of glucocorticoids in mammalian neurons, *Neuroendocrinology,* 53 (Suppl. 1), 25, 1991.

184. **Filipini, D., Gijsbers, K., Birmingham, M.K., and Dubrovsky, B.,** Effects of adrenal steroids and their reduced metabolites on hippocampal long-term potentiation, *J. Steroid Biochem.,* 40, 87, 1991.

185. **Kubli-Garfias, C.,** Chemical structure of corticosteroids and its relationship with their acute induction of lordosis in the female rat, *Horm. Behav.,* 24, 443, 1990.

186. **Hayden-Hixon, D.M. and Ferris, C.F.,** Cortisol exerts site-, context- and dose-dependent effects on agonistic responding in hamsters, *J. Neuroendocrinol.,* 3, 613, 1991.

187. **McGinnis, M.Y., Rutenberg, E., and Lumia, A.R.,** Corticosterone facilitates lordosis behavior and proceptivity in estrogen-primed female rats, *Soc. Neurosci. Abstr.,* 19, 586, 1993.

188. **Witt, D.M., DeVries, A.C., and Carter, C.S.,** Corticosterone modulation of reproductive behavior in prairie voles, *Soc. Neurosci. Abstr.,* 19, 585, 1993.

189. **Moore, F.L. and Miller, L.J.,** Stress-induced inhibition of sexual behavior: corticosterone inhibits courtship behaviors of a male amphibian *(Taricha granulosa), Horm. Behav.,* 18, 400, 1984.

190. **Moore, F.L. and Orchinik, M.,** Multiple molecular actions for steroids in the regulation of reproductive behaviors, *Sem. Neurosci.,* 3, 489, 1991.

191. **Zoeller, R.T. and Moore, F.L.,** Correlation between immunoreactive vastocin in optic tectum and seasonal changes in reproductive behaviors of male rough-skinned newts, *Horm. Behav.,* 20, 148, 1986.

192. **Boyd, S.K. and Moore, F.L.,** Gonadectomy reduces the concentrations of putative receptors for arginine vasotocin in the brain of an amphibian, *Brain Res.,* 541, 193, 1991.

193. **Rose, J.D., Moore, F.L., and Orchinik, M.,** Rapid neurophysiological effects of corticosterone on medullary neurons: relationship to stress-induced suppression of courtship clasping in an amphibian, *Neuroendocrinology,* 57, 815, 1993.

194. **Rose, J.D., Kinnaird, J.R., and Moore, F.L.,** Interactive effects of vasotocin and corticosterone on sensory responses of medullary neurons: implications for stress effects on neuropeptide control of reproductive behavior in an amphibian, *Soc. Neurosci. Abstr.,* 19, 585, 1993.

195. **Boyd, S.K. and Moore, F.L.,** Evidence for GABA involvement in stress-induced inhibition of male amphibian sexual behavior, *Horm. Behav.,* 24, 128, 1990.

196. **Reilly, J.J, Mamini, D.B., Schulkin, J., Slotnik, B., McEwen, B.S., and Sakai, R.R.,** Adrenal steroid implants in the amygdala arouse sodium intake in the rat, *Soc. Neurosci. Abstr.,* 19, 582, 1993.

197. **Thornton, S.N., Martial, F.P., Mousseau, M.-C., Lienard, F., and Nicolaidis, S.,** Rapid action of iontophoretically applied aldosterone on medial septal neuron activity in normal and mineralocorticoid pretreated rats, Submitted, 1994.

198. **Joels, M. and De Kloet, E.R.,** Mineralocorticoid effects on electrical activity in the rat brain, in *Aldosterone: Functional Aspects,* Bonvalet, J.P., Framan, N., Lombes, M., and Rafestin-Oblin, M.E., Eds., John Libbey Eurotext, 1991, 239.

199. **Moore, F.L., Bradford, C.S., and Orchinik, M.,** Distinct high-affinity binding sites for aldosterone and corticosterone on neuronal membranes, *Soc. Neurosci. Abstr.,* 18, 895, 1992.

200. **Plemenitas, A., Lenasi, H., and Hudnik-Plevnik, T.,** Identification of progesterone binding sites in the plasma membrane of the filamentous fungus *Cochliobolus lunatus, J. Ster. Biochem. Mol. Biol.,* 45, 281, 1993.

Chapter 5

RAPID EFFECTS OF ALDOSTERONE:
A NOVEL CONCEPT OF NON-GENOMIC
STEROID ACTION EVOLVES

Martin Wehling

TABLE OF CONTENTS

I. INTRODUCTION

There is an abundance of evidence for delayed, genomic aldosterone effects in renal tubules, its classical target tissue (see Chapters 1–3). Despite the concentration of efforts on the analysis of the mechanisms underlying these processes, extrarenal cardiovascular actions of mineralocorticoids have been demonstrated in nephrectomized dogs as early as 1959.[1] However, comparably little information on the cellular mechanisms involved has been supplied to the scientific community over the past four decades. This may not only reflect the lack of an easily accessible cell model for the study of extrarenal, nonepithelial mineralocorticoid effects, but also the *unquestioned dogma of the genomic pathway* as the only relevant mechanism of steroid action. Over the past few years, latter theory of steroid action is increasingly complemented by accumulating evidence for an additional, completely different mechanism for steroid effects with potential extrarenal, especially cardiovascular, relevance. Relatively early observations in *epithelial* tissues pointed to rapid aldosterone effects in renal cells and rapid, non-genomic steroid effects in the brain, which have been dealt with in chapters 3 and 4. More conclusive evidence for a membrane-bound, novel, aldosterone-specific receptor mainly stems from experiments on very rapid aldosterone effects in *nonepithelial* cells, human mononuclear leukocytes (HML), and rat vascular smooth muscle cells (VSMC) to which this chapter is devoted. Though possibly present in classical target organs of mineralocorticoids as well, the detection of the non-genomic steroid effector mechanism appears to be facilitated in nonepithelial, extrarenal tissues because of the easy access to, and accurate measurement of, transmembrane ion movements in some of these cells *ex vivo*, e.g., white blood cells. These data are summarized here following a short review of extrarenal, *nonepithelial* mineralocorticoid receptors and effects. Their subsequent study resulted in findings clearly incompatible with the dogma of genomic steroid action, and, as a paradigm of scientific serendipity, finally and unexpectedly led to the physiological and pharmacological description of non-genomic aldosterone effects mediated by novel aldosterone-selective membrane receptors. These results are discussed with regard to a new, two-step model of steroid action, and to their clinical relevance.

II. NONEPITHELIAL MECHANISMS OF MINERALOCORTICOID ACTION: EXTRARENAL RECEPTORS AND GENOMIC CARDIOVASCULAR EFFECTS

High affinity, Type-I-mineralocorticoid binding sites have not only been described in classical epithelial target tissues, such as kidney tubules, but also in non-classical tissues such as hippocampus, arterial smooth muscle cells, mammary gland, and human mononuclear leukocytes (HML).[2] Arriza et al.[3] cloned the human mineralocorticoid receptor and demonstrated its mRNA in various tissues such as hippocampus, kidney, heart, spleen, and pituitary. A superfamily of intracellular steroid receptors has now been defined and extensively studied, and the genomic responses are currently being characterized at the molecular level (see Chapters 1 and 2).

Thus, it is evident that these locations not only include epithelial, extrarenal cells (hippocampus, mammary gland), but also nonepithelial, extrarenal cells (arterial smooth muscle cells, HML). Given the numerous locations of high affinity mineralocorticoid binding sites, extrarenal effects of mineralocorticoids were to be expected, and in fact, described before related receptors have been found.

Chronic effects of adrenal steroids on blood pressure, and on electrolyte and water balance, are usually interpreted as a result of epithelial effects in the kidneys, mainly located in the distal tubule.[4] This effector mechanism is considered a genomic action, and involves the binding of aldosterone to cytoplasmic receptors[5] and the transcription of mRNA. From this mRNA, aldosterone-induced proteins (AIP's) are translated,[6] which are thought to create or activate sodium "channels" in the apical cell membrane,[7] so that passive entrance of sodium into the cells is enhanced and the intracellular concentration of sodium passively increases. As a secondary step, the increased intracellular sodium concentration activates and/or induces the production of Na-K-ATPase molecules.[8] Due to the polarity of the membrane distribution of the Na-K-ATPase in tubular cells, transcellular transport of sodium and potassium follows.[9] There are, however, several observations about deoxycorticosterone-acetate (DOCA)-induced hypertension which do not support this renal mechanism as the only cause of mineralocorticoid-induced hypertension.

In DOCA-treated pigs, a marked rise in peripheral resistance was found as a cause for arterial hypertension in addition to increases in extracellular space and cardiac output.[10] Beta blockade of DOCA-treated dogs normalized the cardiac output but did not correct the arterial hypertension.[11] A similar finding was reported for dogs after 16 d of parenteral aldosterone dosage.[12]

In even older studies, doubts about the renal involvement in the development of DOCA-induced hypertension were raised: in bilaterally nephrectomized dogs and rats, arterial hypertension could be induced by DOCA and salt, per se.[1]

Attempting to identify underlying mechanisms, Jones,[13] and Jones and Hart,[14] showed an increase of the membrane permeability for monovalent cations in smooth muscle cells from DOCA-hypertensive rats. By different techniques, increased permeability of sodium, potassium, and lithium was found in smooth muscle cells from DOCA-hypertensive rats compared with cells from normal animals.[15] Kornel et al.[16] showed an increased influx of sodium into smooth muscle cells from the aorta of DOCA-treated rats.

These extrarenal, cardiovascular mineralocorticoid effects are still compatible with the traditional genomic theory of steroid action since long-term hormone actions, rather than acute processes, were studied in later references.

In addition, with regard to ligand affinity and specificity, the pharmacological features of the putative receptors involved resemble those of renal Type I receptors which are commonly thought to be responsible for major mineralocorticoid effects. Their role in human pathophysiology has yet to be determined. In pathological states, such as Addison's or Sheehan's disease, absent gluco- and/or mineralocorticoid effects on the cardiovascular system result in a profound loss of cardiovascular regulation control with low peripheral resistance, heart failure and, finally, death in cardiorespiratory shock.

III. NONEPITHELIAL MECHANISMS OF MINERALOCORTICOID ACTION: NON-GENOMIC ALDOSTERONE EFFECTS MEDIATED BY NOVEL MEMBRANE RECEPTORS

A. VASCULAR SMOOTH MUSCLE CELLS

Aside from rapid renal and neural effects of aldosterone (see Chapters 3 and 4) there is increasing evidence for rapid presumably non-genomic effects of mineralocorticoids in various tissues which are insensitive to the inhibitors of transcription and protein synthesis, actinomycin D, and cycloheximide. The first data tentatively suggesting non-genomic mineralocorticoid effects were reported by Moura and Worcel,[17] who in response to aldosterone demonstrated a late Na-K-ATPase-dependent and an early ouabain-independent efflux of ^{22}Na in the rat tail artery. The ouabain-independent efflux of ^{22}Na was increased as early as 15 min after aldosterone injection (first time point of measurements), while changes of the ouabain-dependent, actinomycin-D-sensitive sodium efflux started after 1 h. Actinomycin D did not alter the rapid responses to aldosterone. These results on smooth muscle cells are consistent with the assumption that aldosterone can increase both entry and exit of Na^+, primarily by enhancement of the plasma membrane "permeability" (for specific mechanisms, see below), and secondarily by increasing the Na-K-ATPase activity. This activation of Na-K-ATPase not only leads to a rise in sodium extrusion, as mentioned above, but also to an increase of intracellular potassium since Na-K-ATPase extrudes sodium in exchange for potassium. The result of both the primary increase of sodium influx and secondary activation of Na-K-ATPase is a concordant increase of sodium and potassium as shown in HML. The rapid increase of sodium flux in response to aldosterone was proposed by these authors as most probably reflecting a non-genomic membrane action of aldosterone. No detailed data on the cellular mechanisms of this effect were available until similar changes of electrolyte transport, including an activation of the sodium-proton-exchanger of the cell membrane, were found in HML and studied extensively (see below).

To further elucidate the mechanisms and the exact time course underlying these 15 min effects of aldosterone in rat tail artery, the effect of aldosterone on the sodium-proton exchanger was investigated in cultured VSMC from rat aortas.[18] The Na^+/H^+-exchanger of the cell membrane is an important membrane transport system for the regulation of intracellular pH and Na^+ in these, as in many other, cells.[19] While rapid activation of the exchanger by mitogens, such as epidermal growth factor, platelet-derived growth factor, and phorbol esters,[20] points to its possible involvement in cell growth, its activation by vasoconstrictors such as angiotensin II or by α-thrombin suggests its regulatory significance for various other functions.[21,22] In addition, the glucocorticoid hydrocortisone at micromolar concentrations stimulates the Na^+/H^+-antiport of VSMC after a latency of at least 4 h by up to tenfold,[23] presumably via the classical genomic pathway of steroid action. Smaller, immediate effects occurring within minutes have not been searched for in this study. Hydrocortisone also induces a phenotypic change of the VSMC and inhibits proliferation of cultured rat aortic VSMC, if proliferation is induced by 10% fetal calf serum (FCS).[23] These delayed actions are likely to involve binding of steroids to classical intracellular receptors, modification of transcription, translation and protein synthesis, and are characterized by a latency of more than 2 hours.

Non-genomic steroid effects on the Na^+/H^+-antiport of cultured rat VSMC were investigated in the range of up to 5 min, thus excluding genomic responses by the time interval chosen.

To minimize the effects of internal Na^+ on the activity of the Na^+/H^+-antiport, rat VSMC were preincubated in an iso-osmotic Na^+/HCO_3^--free buffer at pH 7.0. Under these conditions, internal Na^+ decreases and the cells are acidified to a resting pH of about 6.8. Upon readdition of Na^+, there is a rapid influx of Na^+ which is linear for the first minute and then slightly decreases[22] over the next 5 min.

VSMC were growth arrested with 0.4% FCS from 48 to 24 h before experiments and then incubated without FCS until cells were used to remove steroids present in FCS and establish basal sodium-proton-exchange activity. VSMC remain attached to the tissue plastic dishes and do not alter their morphological appearance under these conditions. Basal $^{22}Na^+$-influx was 22.1 $\pm$ 1.8 nmol /mg protein/min. Aldosterone (1 nM) significantly increased Na^+-influx to 28.6 $\pm$ 1.5 nmol/mg protein/min when added to the preincubation medium only 4 min before the addition of the radiotracer. Na^+-influx after incubation with EIPA (ethylisopropylamiloride, a specific inhibitor of the sodium-proton-exchanger), $\pm$ aldosterone was 9.8 $\pm$ 1.3 and 9.7 $\pm$ 1.1 nmol/mg protein/min; thus the EIPA-sensitive Na^+-influx is significantly (p $<$0.05) increased from 12.3 $\pm$ 1.6 to 18.9 $\pm$ 1.3 nmol/mg protein/min.

Stimulation of EIPA-sensitive Na^+-influx was dose dependent, the half maximal effect (EC_{50}) of aldosterone was seen at ~0.1 nM; fludrocortisone was also active at concentrations between 0.1 and 100 nM with half maximal effects (EC_{50}) at about 0.5 nM. Hydrocortisone did not stimulate EIPA-sensitive Na^+-influx in rat VSMC up to concentrations of 1 μM.

Incubation of VSMC with 1 nM aldosterone, plus 0.1 μM or 1 μM of the classical mineralocorticoid receptor antagonist canrenone, did not result in values different from those after incubation with aldosterone alone, and 0.1 μM and 1 μM canrenone alone were inactive in terms of EIPA-sensitive Na^+-influx.

It is evident that these rapid effects of aldosterone are not explained by the classical theory of steroid action, but rather point to a non-genomic pathway involving membrane receptors. The aldosterone selectivity implied by a very low affinity of cortisol and canrenone is a discriminatory feature of these rapid effects on transmembrane electrolyte transport, separating them from properties of classical genomic steroid actions and Type I mineralocorticoid receptors. The data on aldosterone effects in VSMC presented here share major similarities with the receptor-effector mechanisms for non-genomic aldosterone action in HML. This includes the rapid onset of the stimulation of the EIPA-dependent Na^+-influx, and an effector selectivity for aldosterone which has at least a 1000-fold higher activity than hydrocortisone and canrenone. It is obvious that membrane receptors for aldosterone, but not the classical Type I mineralocorticoid receptors, are ideal candidates for the transmission of rapid aldosterone effects (see below). In addition, the EC_{50} of aldosterone effects in HML and VSMC of ~0.1 nM is close to the physiological concentration of free aldosterone in human (~0.1 nM)[24] and rat plasma (~0.2 nM),[25] thus pointing to a possible physiological cardiovascular relevance of the rapid aldosterone effector mechanism in VSMC.

Though speculative at the moment, it is tempting to assume a regulatory function of the rapid aldosterone effector mechanism in VSMC for the regulation of vascular tone and, thus, peripheral resistance. Immediate postural responses of aldosterone plasma levels would be without effect in a slow-reacting, genomic effector system. However, they could be effective for the peripheral regulation of circulation through an alternative pathway of mineralocorticoid action involving the vascular smooth muscle cells, the main peripheral cardiovascular effector cell.

B. HUMAN MONONUCLEAR LEUKOCYTES

In the search for an easily accessible human cell model to study extrarenal, nonepithelial mineralocorticoid actions, HML were investigated with regard to electrolyte effects of aldosterone. The choice of these cells was triggered by the initial findings of Armanini et al.[2] which described specific binding sites similar to Type-I receptors in these cells. In addition, the study of HML was of primary interest since these resistant cells are well characterized with regard to major determinants of cellular electrolyte and volume regulation. These cells are also available from patients in sufficient amounts and allow the investigation of cellular responses in pathological states of human biology. Paradoxically, those effects of aldosterone found in HML do not predominantly depend on these intracellular binding sites, but different receptors in the plasma membrane (see below).

To address the question of a physiological function for Type I aldosterone binding sites in HML, the effects of aldosterone on electrolyte content, free intracellular calcium, cell volume, and the activity of the sodium-proton exchanger have been examined in these cells *in vitro*. In the absence of aldosterone, the intracellular Na^+, K^+ (Figure 1) and Ca^{2+} concentrations of HML

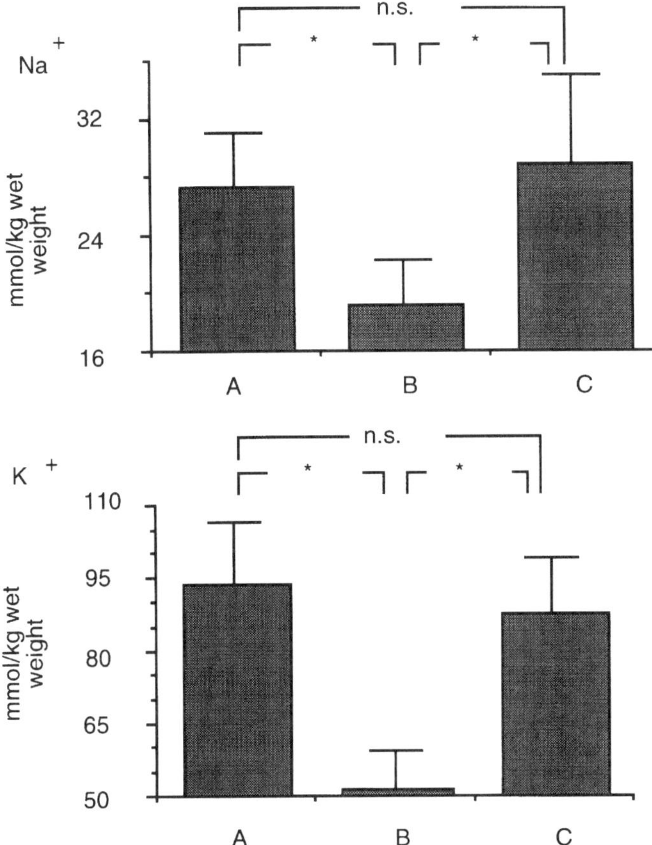

Figure 1. Intracellular Na^+ and K^+ in HML (10 to 20 million cells) before incubation (A), after 1 h incubation at 37°C in RPMI-1640 medium alone (B) or in the presence of 1.4 nM aldosterone (C). Single values and mean + SD are shown. The straight lines join data from the same donor. * p <0.05 compared with B. (Adapted from Wehling, M., Armanini, D., Strasser, T., Weber, P. C., *Am. J. Physiol.*, 252, E505, 1987. With permission.)

decrease significantly within 1 h if HML are incubated in RPMI-1640 at 37°C, whereas these concentrations remain constant when aldosterone is added to the incubation medium. These changes reflect the wash-out of the aldosterone effects exerted by physiological concentrations of the steroid still present in freshly isolated HML. This wash-out is prevented by the addition of aldosterone to the incubation medium *in vitro*. The effects are half maximal at aldosterone concentrations of approximately 0.1 nmol/l, and are accompanied by parallel shifts of cell water and, thus, cell volume (Figure 2) by ~15%.[26-28] The classical mineralocorticoid antagonist canrenone at least partially antagonized these 1 h effects of aldosterone.

Concordant changes in intracellular sodium, potassium, calcium, and cell volume thus identify aldosterone as a major physiological determinant in the

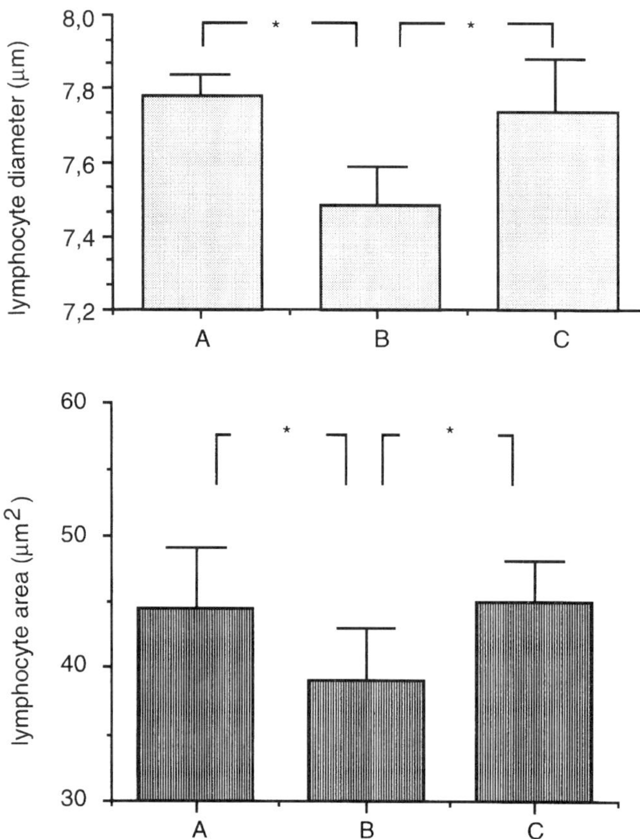

Figure 2. Diameters (upper panel) and areas (lower panel) of 50 to 100 human mononuclear leukocytes from 6 normals were measured before (A) and after 1 h incubation in RPMI-1640 medium at 37°C without (B) and with 1.4 n*M* aldosterone added (C). Means ± SD and values for individual probands are depicted. Values for single donors are connected by straight lines. * *p* <0.05. (Adapted from Wehling, M., Kuhls, S., Armanini, D., *Am. J. Physiol.*, 257, E170, 1989. With permission.)

regulation of lymphocyte volume in isotonic media. A similar concordant change of sodium, potassium, and cell volume has been demonstrated for the regulatory response of HML to anisotonic media. If exposed to hypertonic media, HML shrink rapidly and then regain volume (regulatory volume increment, RVI), mainly by activation of the sodium-proton exchanger in the cell membrane and subsequent amiloride-sensitive sodium transport.[29] This cation transport is coupled with a bicarbonate-chloride exchange to balance the intracellular alkalinization, so that sodium chloride enters the cell in RVI. This gain of sodium is paralleled by an increase in intracellular potassium by an activation of the Na-K-ATPase.[29] The free intracellular calcium is commonly considered a second messenger for triggering and regulation of contractile force, which may in turn be influenced by intracellular sodium via a Na^+/Ca^{2+}-exchanger in the plasma membrane.[30]

From the results cited above, the following sequence of mineralocorticoid effects on HML electrolytes and volume can be postulated: aldosterone facilitates the entrance of sodium into HML, thus increasing the intracellular sodium concentration. This activates the Na-K-ATPase, which exchanges part of the sodium for potassium. A net gain of both sodium and potassium results in an increased cellular osmolarity and, thereby, in cell swelling by water uptake. The free intracellular calcium is coupled to the intracellular sodium by a putative Na^+/Ca^{2+}-exchanger of the cell membrane of HML.[31]

To elucidate the primary events by which aldosterone facilitates sodium entry into HML we have focused on effects of aldosterone on the sodium-proton exchanger in HML since this membrane transport system is a common target of various hormones, including growth factors and angiotensin II, and involved in RVI in HML as described above. In HML, the activity of the sodium-proton exchanger is conveniently measured electronically in a Coulter Counter® by the rate of EIPA-sensitive swelling in an isotonic sodium propionate medium.[32] A significant stimulation of this transporter can be seen starting as early as 1 to 2 min after incubation with aldosterone, with an EC_{50} for aldosterone of 0.04 nM (Figures 3 and 4). The effect is antagonized by amiloride and EIPA, the amiloride-analogue which is commonly considered as a specific inhibitor of the sodium-proton exchanger, but not by actinomycin D or cycloheximide. Cortisol and dexamethasone are weak agonists requiring concentrations 4 orders of magnitude higher than that of aldosterone for an even smaller effect (Figure 5). The classical mineralocorticoid receptor antagonist canrenone did not antagonize this effect of aldosterone at concentrations of up to 1000-fold that of aldosterone. Canrenone itself was inactive.[33,34]

Thus, the effects of aldosterone on electrolytes and cell volume in HML appear to be comparable to those initiated by the RVI as mentioned above: aldosterone stimulates the sodium-proton exchanger, thereby increasing the intracellular sodium concentration. This in turn initiates the cascade of events (Figure 6) discussed above (activation of the Na-K-ATPase, cell gain in potassium, water, increased volume, and free intracellular calcium).

Certain characteristics of the effect of aldosterone on the sodium-proton exchanger in HML become clear from the present data with regard to the location and types of receptors involved. The data suggest that these receptors could be different from the known mineralocorticoid receptors and transmit non-genomic responses of HML to mineralocorticoids. The evidence, though not proof, for this derives from the following findings:

1. The initial step of electrolyte effects, the activation of the sodium-proton exchanger of the cell membrane, begins within 1 to 2 min after application of mineralocorticoids. These time intervals are commonly considered to be too short for the complete activation or shut off of DNA-dependent processes, including binding to cytosolic receptors, binding of

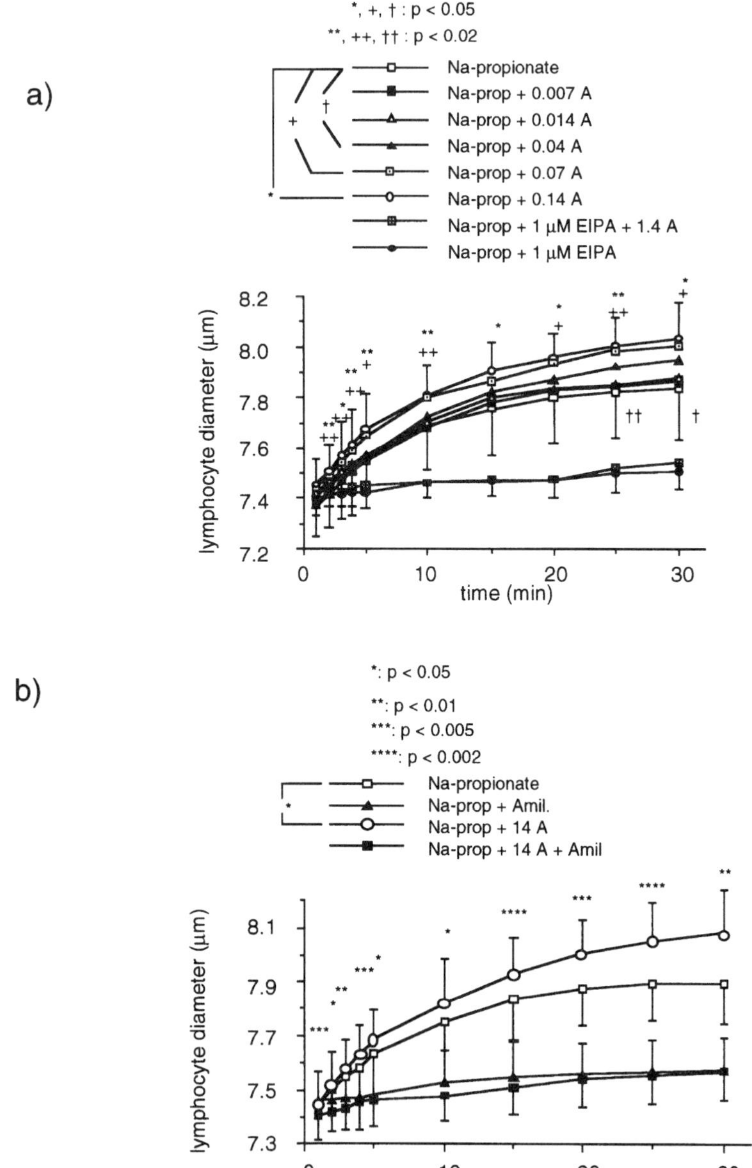

Figure 3. The peak diameters of human mononuclear leukocytes obtained from the diameter distribution curve are shown during incubation in isotonic Na^+-propionate buffer alone, or with 0.007 to 1.4 nM aldosterone and/or 1 µM ethylisopropylamiloride (EIPA) were (a) added. Figure b shows the effect of Na^+-propionate buffer alone, or with 14 nM aldosterone, 400 µM amiloride or 14 nM aldosterone plus 400 µM amiloride on HML swelling. The incubation was done at room temperature, values are means ± SD for 7 to 8 (a) or 14 (b) experiments from different normal donors. A: aldosterone; Amil: amiloride. (Adapted from Wehling, M., Käsmayr, J., Theisen, K., *Am. J. Physiol.*, 260, E719, 1991. With permission.)

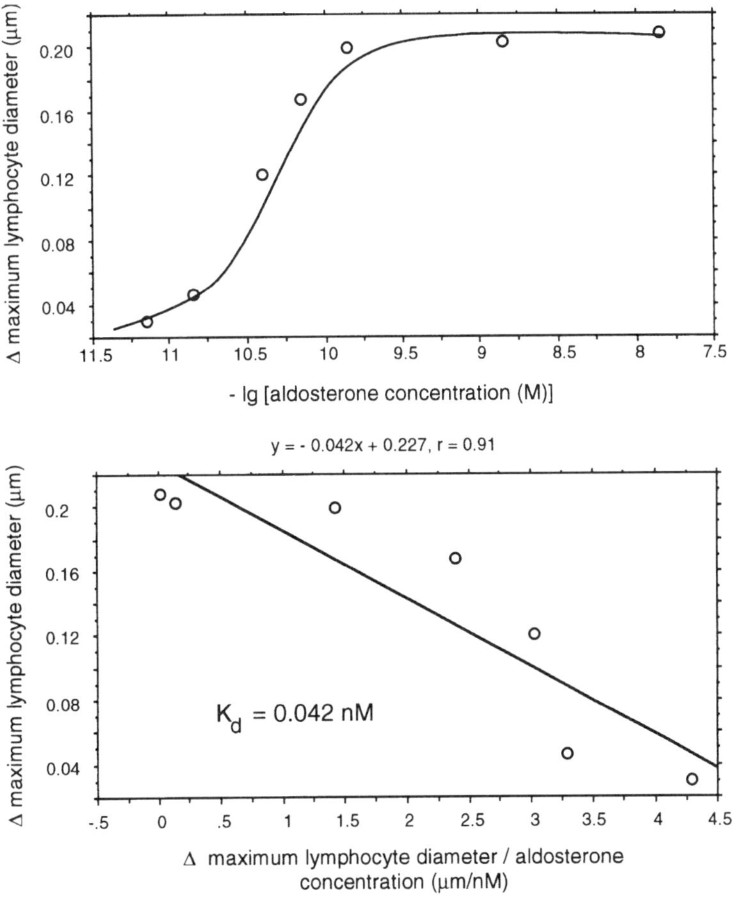

Figure 4.   The dose-response curve for maximum effects of aldosterone on the HML diameter after incubation for 30 min in Na^+-propionate is shown in the upper panel (mean of 7 to 14 experiments). The lower panel shows the Scatchard plot for the maximum change of the HML diameter by aldosterone (0.007 to 14 n*M*). The equation for the linear regression is shown above. (Adapted from Wehling, M., Käsmayr, J., Theisen, K., *Am. J. Physiol.*, 260, E719, 1991. With permission.)

the complexes to the nuclear DNA, production of m-RNA, translation to proteins, eventually modification, e.g., glycosylation of proteins, transport of proteins to the membrane, insertion of proteins into the plasma membrane and, finally, subsequent activation of electrolyte transport systems. In cultured mammalian collecting duct cells, synthesis of AIP's starts after 1 h.[35] As the fastest genomic event known in the action of mineralocorticoids, the transcription of mouse mammary tumor virus long terminal repeat (MMTV LTR), begins within 30 min and reaches a plateau after 3 h in a feline renal cell line.[36] This process, which is

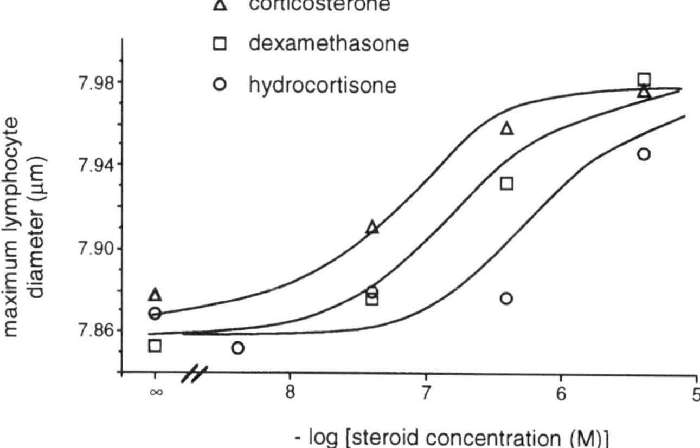

Figure 5. The peak diameters of human mononuclear leukocytes obtained from the diameter distribution curve are shown during incubation in isotonic Na^+-propionate with or without different steroids added as indicated. Values are means $\pm$ SD for 7 to 14 experiments from different normal donors. The dose-response curves of hydrocortisone, corticosterone and dexamethasone are shown. (Adapted from Wehling, M., Käsmayr, J., Theisen, K., *Am. J. Physiol.*, 260, E719, 1991. With permission.)

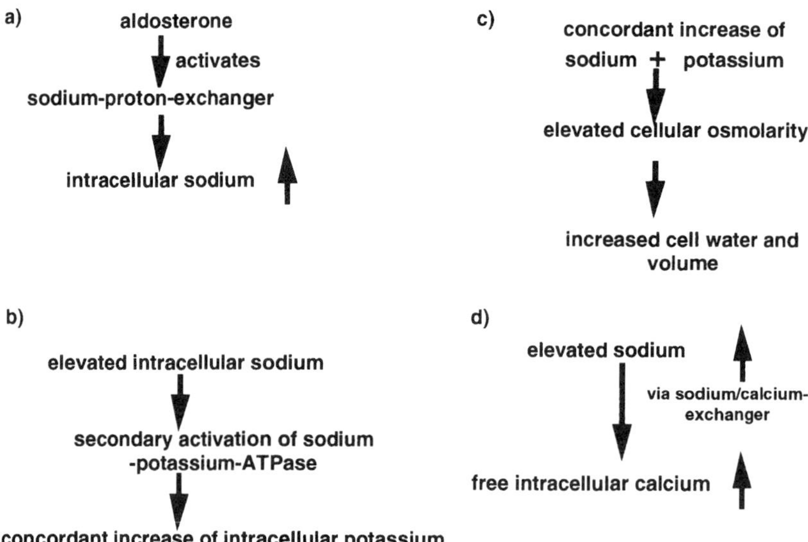

Figure 6. The sequence of events for the influence of aldosterone on electrolyte and water transport as shown in HML. For details see text.

    insensitive to cycloheximide, does not even include a subsequent synthesis of protein. For glucocorticoids, effects on MMTV LTR in L tk⁻ aprt⁻ cells were first seen after 7.5 min.[37] To our knowledge, no complete sequence of genomic action, including protein synthesis, is known for mineralocorticoids which is detectable after 1 to 2 min at room temperature.

2. The K_d of the binding of aldosterone to intact HML is about 1.4 nM, the EC_{50} for the effects on the sodium-proton exchanger, cell electrolytes, and volume is about 0.1 nM. This difference of more than one order of magnitude between receptor binding and the electrolyte effects strongly suggests the existence of two different types of mineralocorticoid binding sites. Earlier studies on aldosterone binding to toad bladder also showed two types of mineralocorticoid receptors with K_d values of 0.3 and 50 nM.[38] An early response of transepithelial fluxes (0.75 to 2.5 h) was related to the high affinity binding site and a late response (>6 h) to the low affinity binding site. The large differences in the concentration ranges of aldosterone and hydrocortisone which affect the sodium-proton exchanger also suggest that the receptors involved are different from the cloned Type I mineralocorticoid receptor which does not discriminate between both steroids.[3] In addition, actinomycin D and cycloheximide did not inhibit the activation of the sodium-proton exchanger by aldosterone to a major extent in HML.

These arguments, especially the rapid onset of the activation of the sodium-proton exchanger by aldosterone, suggest that these effects of aldosterone are transmitted by a non-genomic pathway by receptors different from those intracellular, Type I receptors which are involved in genomic steroid actions and have been demonstrated in HML by Armanini et al.[2] These receptors should have properties differing uniquely from those of the classical Type I receptors: their selectivity for aldosterone over cortisol and canrenone appears to represent a landmark for their biochemical identification.

Not surprisingly, in HML, two types of aldosterone-binding receptors were described in agreement with this concept of the coexistence of both intracellular Type I receptors and membrane receptors for mineralocorticoids:

1. Intracellular Type-I Mineralocorticoid Receptors

Armanini et al.[2] first reported specific ³H-aldosterone binding to intact HML with an apparent affinity constant (K_d) of 1.4 nM. Canrenone inhibited aldosterone binding, the affinity for cortisol was one fifth of that for aldosterone. Specific binding to HML separated from peripheral blood by Percoll® was measured in the presence of a 5000-fold excess of the synthetic, "pure" glucocorticoid RU 26988. This compound restricts the binding of aldosterone to mineralocorticoid receptors and thus allows the selective study of these receptors. The affinity constant of aldosterone binding to HML mineralocorticoid receptors was in the supraphysiological range of the free plasma aldosterone concentration ($\sim$0.1 nM), the number of receptors/cell was 200 to 400. No aldosterone binding to platelets and granulocytes was observed. Thus, the receptor described by Armanini resembles classical Type I receptors transmitting genomic responses of mineralocorticoids in many regards. Arriza cloned this classical intracellular Type I receptor from human kidney[3] and showed binding of aldosterone and cortisol at a ratio of 1:1, and of aldosterone and canrenone at a ratio of 1:5. The K_d for aldosterone binding was 1.3 nM.

In contrast to its role in classical target tissues, the function of Type I mineralocorticoid receptors in HML is not yet completely understood; their involvement in 1 h effects of aldosterone on HML Na^+, K^+ and cell volume may be hypothesized from the action of canrenone, which was partially active as an inhibitor for these particular aldosterone effects only. As mentioned above, canrenone was inactive as an inhibitor for the rapid effects of aldosterone on the sodium-proton exchanger, indicating the insignificance of these receptors for the initial, immediate steps of aldosterone action in HML. After 1 h of incubation with aldosterone, a genomic response, including an increased synthesis of Na-K-ATPase molecules, may be initiated via classical Type I receptors, and *augment* the effects of primary rapid actions of the steroid, resulting in changes of total cell sodium, potassium, and water content.

2. Membrane Receptors for Aldosterone

a. Binding Experiments

In the traditional theory on steroid hormone action, the mineralocorticoid effector mechanism involves binding of aldosterone to cytosolic and/or nuclear Type I receptors with subsequent synthesis of specific proteins (AIP's). In contrast, rapid *in vitro* effects of aldosterone on the activity of the sodium-proton exchanger in HML,[33,34] VSMC,[17] and kidney cells[39] have been demonstrated which are incompatible with the involvement of these classic steroid binding sites. They suggest the existence of distinct membrane receptors which are currently of expanding interest with regard to steroids other than aldosterone. This includes the effects of 5α-pregnan-3α,21-diol-20-one and 5α–pregnan-3α-OH-20-one on the $GABA_A$-receptor, of pregnanolone on luteinizing hormone releasing peptide (LHRH) secretion, of progesterone on dopamine release, neural effects by the local application of steroids, the action of progesterone on the oocyte maturation, spermatozoan acrosome reaction, and a variety of other steroid effects and/or membrane binding sites. From these data, the existence and involvement of membrane receptors appears as a general principle in steroid hormone action. This principle might be ubiquitous, including the action of noncortical steroids such as vitamin D_3 and other hormones which are thought to primarily act through protein synthesis, e.g., thyroid hormones.

Studies were performed to identify putative plasma membrane receptors for aldosterone which could be responsible for the rapid actinomycin D and cycloheximide insensitive stimulation of the Na^+/H^+-antiporter of HML.[33,34]

HML plasma membranes were prepared via nitrogen cavitation and differential centrifugation. In the radioassay developed for that study, an iodinated analogue of aldosterone, aldosterone-3-(O-carboxymethyl)-oximino-(2-[125I]iodohistamine), was used as tracer. The iodinated radioligand had to be used to allow the identification of a binding site with an expected K_d-value of ~0.1 nM with sufficient accuracy. The commonly used tritiated aldosterone has a specific activity that is too low to achieve this. The extensive, nonspecific binding of the lipophilic radioligand aldosterone-3-(O-carboxymethyl)-oximino-(2-[125I]iodohistamine) to protein and the filters had to be addressed by pilot testing of different filters and washing procedures. Since the specific binding to mineralocorticoid membrane receptors had a high turnover (see below), the washing time in the filter assay was crucial: it had to be short enough to warrant the detection of short-life receptor interaction and had to sufficiently reduce nonspecific binding. Glass fiber filters (GF/B and C, Whatman) were used in binding assays because they allowed a rapid washing with large volumes to minimize unspecific filter binding.

The binding of the iodinated aldosterone analogue to HML plasma membranes exposed a saturable feature of its specific (total - nonspecific) binding. The total, specific and unspecific binding, and the Scatchard analysis of this specific binding are shown in Figures 7 and 8 indicating a maximum binding of 34 cpm/µg membrane protein (~10 fmol/mg protein) at a calculated K_d of 0.16 nM for the radioligand. The maximum binding corresponds to a number of 100 to 200 binding sites/lymphocyte.

Analogous experiments with cortisol-3-(*O*-carboxymethyl)-oximino-(2-[125I]-iodohistamine) from 0.01 to 1 nM did not show specific binding, but linear total binding comparable to the nonspecific binding of the iodinated aldosterone analogue (Figure 7). This was done to demonstrate that the steroid moiety, not the histamine-containing side chain which is the same for both radiotracers used, is responsible for specific binding. An analogous Scatchard analysis was obtained for binding of aldosterone-3-(*O*-carboxymethyl)-oximino-(2-[125I]iodohistamine) to plasma membranes from pig kidney (Figure 8).

The displacement effects of various steroidal and nonsteroidal agents in the binding assay identify a number of compounds (cortisol, canrenone, dexamethasone, RU 26988, corticosterone, 18-OH-progesterone, ouabain, and amiloride) which were inactive at physiologically or pharmacologically relevant concentrations up to 1 µM. Aside from aldosterone (K_d calculated by the division of the 50% displacement concentration of aldosterone by the concentration of tracer used times the K_d of the radioligand to be ~0.1 nM),[40] only DOCA exposed a displacing activity (K_d ~100 nM)[41] at concentrations in a range of physiological or pharmacological relevance. The other corticosteroids tested started to displace the tracer only at much higher concentrations (0.01 to 0.1 mM). This is likely due to a nonspecific action as also suggested by an almost uniform concentration at which displacement finally occurs (Figures 9 and 10).

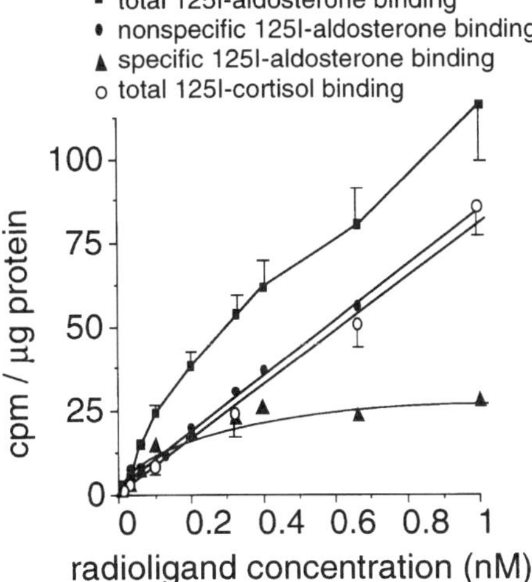

Figure 7. The total, specific and unspecific binding of [125I]-labeled aldosterone and the total binding of [125I]-labeled cortisol to plasma membranes from human mononuclear leukocytes is shown. Plasma membranes were incubated for one hour at 37°C in a physiological buffer (140 mM NaCl, 5 mM KCl, 1 mM CaCl$_2$, 0.5 mM MgCl$_2$ 5 mM glucose, 1 mM Na$_2$HPO$_4$ and 10 mM Tris/ HCl, pH 7.4). Radioligands were added at concentrations from 0.01 to 1 nM. Non-specific aldosterone binding was measured at an excess of 1 µM "cold" aldosterone under otherwise identical conditions and subtracted from total binding to result in specific binding. Means ± SD for 6 experiments are given. (Adapted from Wehling, M., Christ, M., Theisen, K., *Am. J. Physiol.*, 263, E974, 1992. With permission.)

In the case of plasma membranes from pig kidney, 50% of the tracer was displaced by cold aldosterone at a concentration of ~0.1 nM and by cortisol at a concentration of about 1 µM (not shown).

The kinetics of aldosterone binding to HML plasma membranes suggest a rapid turnover of the ligand. The association and dissociation is half maximal as early as 1 min after start of incubation with the tracer, or the chase by cold steroid, and is almost complete after 5 to 10 min.

These findings are the first to demonstrate receptors in the membrane with a high affinity for aldosterone, but not cortisol and canrenone. From the stimulatory effect of aldosterone on the sodium-proton exchanger in HML (see above), the expected K$_d$ value of the putative membrane receptors was estimated to be ~0.1 nM.[41] Tritiated aldosterone is not capable of detecting binding sites in this concentration range unless large amounts of binding material (plasma membranes) are applied. It, therefore, was essential for the acquisition of reliable data to utilize an iodinated analogue of aldosterone, aldosterone-3-(*O*-carboxymethyl)-oximino-(2-[125I]iodohistamine), with a high specific activity of 2000 Ci/mmol. As no binding data for native aldosterone are available, the hypothesis for the data interpretation is based on binding characteristics of the iodinated analogue comparable to those of native aldosterone. This assumption is supported by the following:

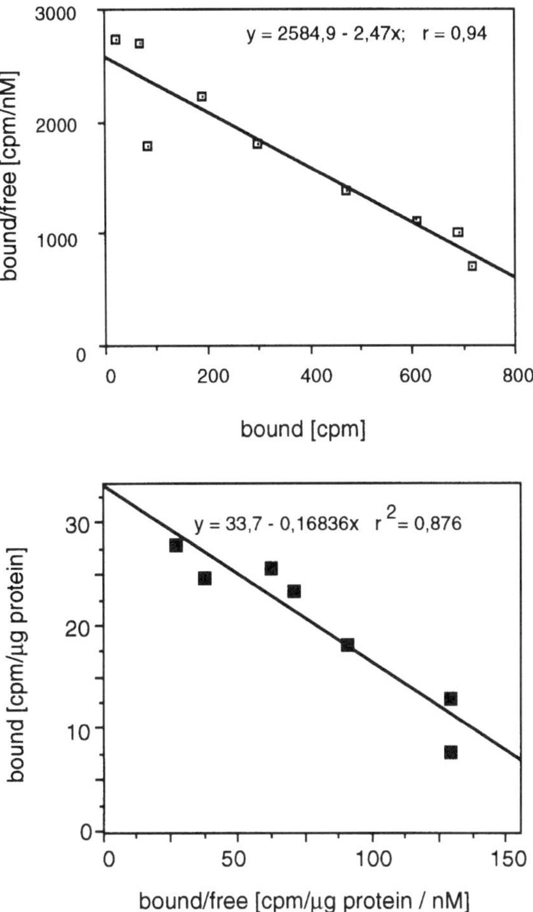

Figure 8.   The Scatchard analysis of the specific binding of aldosterone-3-(*O*-carboxymethyl)-oximino-(2-[125I]iodohistamine) to plasma membranes from pig kidneys (upper panel) and HML. (Lower panel, adapted from Wehling, M., Christ, M., Theisen, K., *Am. J. Physiol.*, 263, E974, 1992. With permission.)

1. The displacement by cold aldosterone results in a similar K_d value for both the tracer and cold aldosterone, with a curve-shape which suggests a first order receptor-ligand interaction.

2. Cortisol, with the same side chain to attach the radioactive iodine containing histamine, did not expose specific binding in excess to nonspecific binding in the concentration range of interest. This identifies the steroid moiety to be essential for the binding properties in this assay.

3. The aldosterone tracer used was characterized for, and used in, radioimmunoassays for aldosterone. Its selectivity for displacement by cold aldosterone in the presence of all other steroids in human plasma has been accepted for clinical and scientific purposes.

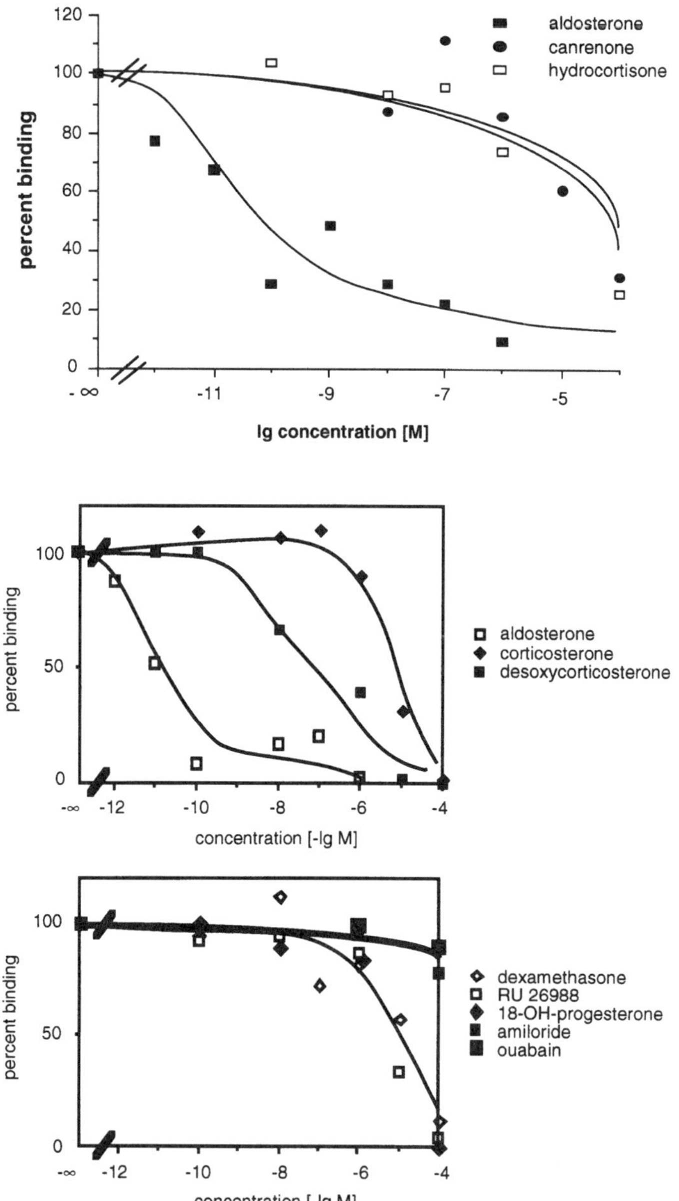

Figures 9 and 10.   The displacement of the radioligand aldosterone-3-(*O*-carboxymethyl)-oximino-(2-[^{125}I]-iodohistamine) by different steroids and non-steroidal compounds from plasma membrane preparations of human mononuclear leukocytes is shown for a radioligand concentration of 0.2 n*M*. Values represent means of 6 to 8 experiments. Curves are drawn by hand. (Figure 9 adapted from Wehling, M., Christ, M., Theisen, K., *Biochem. Biophys. Res. Comm.*, 181, 1306, 1991. With permission. Figure 10 adapted from Wehling, M., Christ, M., Theisen, K., *Am. J. Physiol.*, 263, E974, 1992. With permission.)

An important additional prerequisite for a successful demonstration of the putative membrane receptors was a strategy to purify plasma membranes from HML[42] as binding substrate, which had to achieve two major goals:

1. To remove those cell constituents which are considered as sites of the classical mineralocorticoid receptor, nuclei and cytosol.
2. To generate sufficient amounts of plasma membranes from HML which are the same source as studied in the functional experiments.

The binding substrate used in this study was free of intact cells and nuclei, as shown by light microscopy. DNA content and lactate dehydrogenase as markers for nuclear fragments and cytoplasm, the sites of classical Type I receptors, were reduced by 76 and 79%. Thus, purification from latter cell constituents was not perfect, and small amounts of the cytosolic Type I receptor may still be present. However, all attempts to further purify the binding substrate according to Schmidt-Ullrich et al.[42] resulted in an obstructively low yield of membranes from given amounts of human venous blood. On the other hand, pharmacological properties of the putative membrane receptors were definitely different from that of the cytosolic Type I receptor. Therefore, a clear discrimination was obtained by the characterization of the pharmacological profiles for specific binding without further steps of purification.

Another way to demonstrate aldosterone-specific receptors in the plasma membrane would be the autoradiographic *in situ* localization of aldosterone binding. An analogous approach to localize corticosteroid receptors in neuronal membranes[43] was eased by the polarity of neuronal cells concentrating perikarya and neuropil, and thus nuclei and synaptic membranes in different macroscopic areas. Obviously, lymphocytes are apolar cells and single cell autoradiography would be required. The calculated 100 to 200 binding sites per lymphocyte with an even lower number of radiophotons emitted per cell, and the fast turnover of the interaction seem to definitely obscure the *in situ* localization of membrane receptors for principal reasons, at least in HML. Studies on aldosterone-specific membrane receptors in other tissues might be a clue for this approach of receptor localization (see results for binding to plasma membranes from pig kidneys below), but a richer source of membrane receptors in polar tissues still needs to be found.

Bearing these true limitations in mind, the major arguments against cytosolic or nuclear impurities in the membrane preparation as reason for the aldosterone binding described here remain those binding properties described above which are incompatible with all major aspects of the classical Type I mineralocorticoid receptor.

All effects of aldosterone on HML electrolyte transport and volume, including the stimulatory effect on HML sodium, potassium, and calcium concentrations, the sodium-proton exchanger and cell volume[26-28,33,34] were half maximal at about 0.1 nM. These EC_{50}-values thus perfectly match the K_d of aldosterone binding to plasma membranes from the same cells as reported here. This congruity represents a major prerequisite for the assumption of a causal relation between these effects and the binding sites in the plasma membranes from HML. In addition, a mineralocorticoid receptor in the affinity range of ~0.1 nM would explain the biological activity of physiological plasma concentrations of free aldosterone (in humans ~0.1 nM, in rats ~0.2 nM) rather than a receptor with a K_d of above 1 nM such as the cloned renal mineralocorticoid receptor (K_d 1.3 nM)[3] and the mineralocorticoid receptor in intact HML (K_d 1.4 nM).[2]

In the literature, there are only two earlier reports on mineralocorticoid binding to plasma membranes. The K_d for aldosterone binding was 3 nM for solubilized and 13 nM for intact plasma membranes in one,[44] and 100 nM in another preparation[45] from rat kidneys. The physiological significance of binding sites in this affinity range being far from the physiological concentration of free aldosterone (~0.1 nM) is questionable. In a later reference,[45] pharmacological properties of both membrane and nuclear receptors were similar and an attachment of the classical, intracellular receptor to the cell membrane, or an impurity of the preparation with this receptor, has to be discussed.

As intact HML contain all cell constituents which are thought to harbor the classical mineralocorticoid receptors, namely nuclei and cytoplasm, the binding of tritiated aldosterone to intact HML demonstrated by Armanini et al.[2] very likely included classical Type I mineralocorticoid receptors. Additionally, the pharmacological characteristics of intact HML receptors resembled classical, Type I receptors in many regards: binding affinity of aldosterone and cortisol was not very different, and the classical mineralocorticoid antagonist canrenone also was a high affinity ligand.

The binding features of aldosterone to HML membranes share further important analogies with the characteristics of rapid aldosterone effects on HML electrolytes:

1. The putative membrane receptors are selective for aldosterone and bind neither canrenone nor cortisol significantly at concentrations as high as 0.1 μM,
2. The association and dissociation kinetics of aldosterone binding to HML plasma membranes indicate a rapid turnover of the ligand.

Again, these data perfectly match the functional results of aldosterone effects on the sodium-proton exchanger that are essentially characterized by a triad of characteristic features: onset times of only 1 to 2 min, a minor response to cortisol as agonist at high concentrations only (~1 μM), and no effect of canrenone as antagonist.

The pharmacological properties (K_d- or EC_{50}-values, selectivity for aldosterone over cortisol, and canrenone) of HML membrane binding, activation of the sodium-protonexchanger in HML and VSMC, and stimulation of IP_3 production are summarized and compared with those of the cloned classical Type I mineralocorticoid receptor in Table 1; it is quite obvious that these

TABLE 1

**Comparison of Properties of the "Classic" Cloned Mineralocorticoid
Receptor (CCMC), of the Aldosterone Receptor of the Cell
Membrane (AMC) and of Steroid Effects on the Sodium-Proton-
Antiport (Na/H) or IP$_3$-Production in Human Mononuclear
Lymphocytes (HML) and on the Sodium-Proton-Antiport (Na/H)
in Vascular Smooth Muscle Cells (VSMC)**

| | **Mineralocorticoid Receptors** | | **Rapid Effects of Aldosterone** | |
	CCMC	**AMC**	**Na$^+$/H$^+$-exchanger in lymphocytes/ smooth muscle cells**	**Stimulation of IP$_3$-production in lymphocytes**
K$_d$ or EC$_{50}$				
[nM] for aldosterone	1.3	~0.1 nM	~0.1 nM	~0.1 nM
Aldosterone/				
cortisol affinity	1:1	~10000:1	≥10000:1	~1000:1
Aldosterone/				
canrenone affinity	~5:1	>1000:1	>1000:1	≥100:1

(Adapted from Wehling, M., Christ, M., Gerzer, R., *Trends Pharmacol. Sci.*, 14, 1, 1993. With permission.)

membrane binding sites are ideal candidates to be the receptor which mediates rapid effects of aldosterone in HML.

b. Molecular Structure of Aldosterone Membrane Receptors

The data for specific aldosterone-binding sites in HML plasma membranes, and the perfect match with functional studies in the same cells, are very suggestive evidence for the existence of a distinct receptor protein in the membrane with unique properties of highly selective steroid specificity. On the way to the analysis of its molecular structure, basic characteristics of the receptor protein were obtained by the conventional means of molecular weight determination and protein chemistry. Binding of aldosterone-3-(*O*-carboxymethyl)-oximino-(2-[125I]iodohistamine), the same ligand as used in the binding studies described above, to plasma membranes from HML was investigated on sodium dodecyl sulfate polyacrylamide gel electrophoresis (SDS-PAGE) after photoaffinity labeling with BASED (bis-[β-(4-azidosalizylamido)ethyl]disulfide) in the presence and absence of steroids. Figure 11 shows the radioactivity in 3-mm-slices from SDS-PAGE of plasma membranes from HML which were labeled by BASED with 1 nM aldosterone-3-(*O*-carboxymethyl)-oximino-(2-[125I]iodohistamine) in the absence or presence of a 1000-fold excess concentration of "cold" aldosterone or cortisol (1 μM). The major peak of radioactivity is seen in a slice corresponding to a molecular weight (M$_w$) of ~50 kDa, two minor peaks correspond to a M_w of ~100 kDa and ~200 kDa. 1 μM cold aldosterone completely inhibited labeling at the major, and partially at the minor, peaks. The most importing finding in these experiments concerns the effect of cortisol, which was unable to antagonize labeling by aldosterone at a concentration of 1 μM. This is clearly visible for the 50 kDa peak, and may be suggested from the 100 and 200 kDa peaks (Figure 11). Spreading of the gels above a M_w of ~85 kDa by extended runs for 4 h instead of 2 h more clearly demonstrated the peaks at M_w ~100 and 200 kDa being sensitive to 1 μM aldosterone, but not to 1 μM cortisol. Tracer displacement by aldosterone is only partial at these peaks, indicating a high amount of nonspecific binding. The sulfhydryl agent DTT nonspecifically reduces background binding of the tracer. The peaks at 50, 100, and 200 kDa were not essentially affected by this pretreatment. High-salt solubilization (1 M NaCl) did not result in detectable aldosterone binding in the supernatant.[46]

A high aldosterone selectivity was found in both the functional and membrane binding experiments by the demonstration of an at least 10^3 to 10^4 fold lower affinity of cortisol compared with that of aldosterone.

This aldosterone selectivity is unique for the aldosterone membrane receptor and thus served as a landmark for its identification on SDS-PAGE. The main findings of the study on the analysis of the aldosterone membrane receptor by SDS-PAGE may be summarized as follows:

1. At a M_w of ~50 kDa, there is binding of radiolabeled aldosterone to plasma membrane proteins from HML on SDS-PAGE which is displaced by 1 μM aldosterone, but not by 1 μM cortisol. Two smaller peaks of cortisol-insensitive aldosterone binding at ~100 and ~200 kDa could be identified.
2. The sulfhydryl agent DTT does not reduce specific aldosterone binding indicating the absence of SH-groups in the binding domain or sensitive structures of the receptor.
3. High salt concentrations did not solubilize the receptor protein which, thus, appears to be an integral membrane protein.

A major problem with the identification of steroid receptors in plasma membranes is their low concentration and high turnover of steroidal ligands (see above). Therefore, a technique to covalently label the receptor by its ligand had to be employed to allow the detection of stable receptor-ligand complexes under the aggressive (high concentrations of sodium dodecyl sulfate) conditions of SDS-PAGE. To achieve this, photoaffinity labeling has been widely used with regard to other steroids. In our experiments, BASED was the cross-linking agent which was able to sufficiently label the plasma membrane binding sites for aldosterone in nonreducing SDS-PAGE at one major and two minor peaks. The stochiometry of the molecular weights of the proteins in these peaks strongly suggests the existence of a dimer and a tetramer of the monomeric receptor protein which is found in the 50-kDa peak. A major

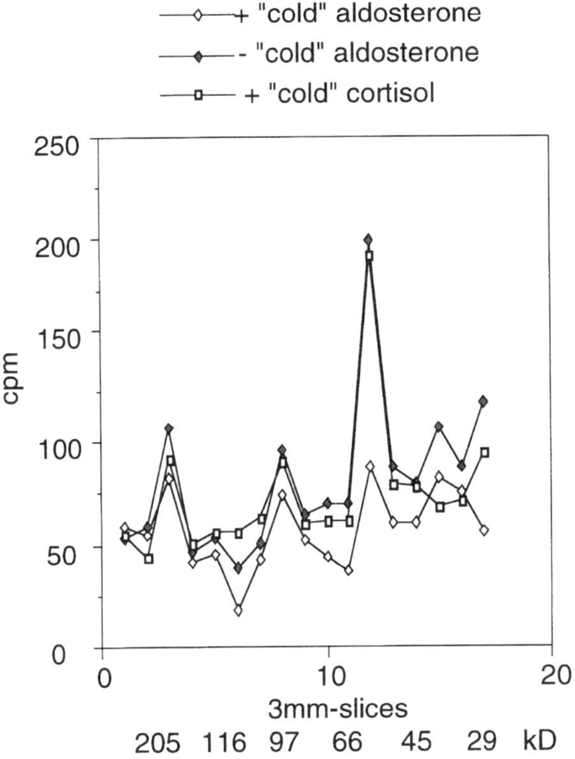

Figure 11. The binding of 1 nM aldosterone-3-(O-carboxymethyl)-oximino-(2-[125I]iodohistamine) in the presence or absence of 1 μM unlabeled aldosterone or cortisol to plasma membrane preparations from human nuclear leukocytes is analyzed after cross-linking with BASED (3 nM) by 7.5% SDS-PAGE. Gels were cut into 3 mm slices before counting. Values represent means of 5 experiments. (Adapted from Wehling, M., Eisen, C., Aktas, J., Christ, M., Theisen, K., *Biochem. Biophys. Res. Commun.*, 189, 1424, 1992. With permission.)

contamination by the the classical cytosolic Type I mineralocorticoid receptor is unlikely for the 50-kDa peak since cortisol at 1 μM was completely ineffective in displacing the tracer, but should displace aldosterone from the classical Type I receptor. However, it had to be suspected in the right hand part of the broad 100-kDa peak since both cortisol and aldosterone were able to displace the radiotracer here.

The identification of the newly described aldosterone membrane receptor as a 50-kDa protein possibly forming dimers and tetramers appears as an important step in the process of the molecular characterization of the receptor protein. It may allow its further purification, concentration, and, finally, sequencing, gene isolation, and cloning. The description of an alternative, nongenomic pathway of mineralocorticoid action seems to include the identification of the completely new class of steroidal membrane receptors of which no member has yet been cloned. The molecular weight of the aldosterone membrane receptor appears to be low in comparison with other membrane receptors, e.g., catecholamine receptors which commonly are found to be in the range of 80000 to 130000 Dalton. However, the question of a composite receptor built from multiple copies of the same or slightly different subunits still has to be settled.

3. Second Messengers for Rapid Aldosterone Effects
a. Inositoltrisphosphate

Although rapid increases of intracellular calcium in oocytes and spermatozoa after progesterone[47,48] application have been demonstrated, little is known about the involvement of other second-messenger systems in fast steroid action. Changes of intracellular inositol-1,4,5-trisphosphate (IP_3) levels occurring in VSMC 15 min after application of hydrocortisone[49] and the modulation of angiotensin II and vasopressin induced IP_3 production in VSMC[50] after glucocorticoid treatment were suggestive for a possible involvement of the IP_3 intracellular signaling system in rapid steroid action. In addition, in response to agents such as growth hormones or angiotensin II, the Na^+/H^+-antiport is stimulated via calcium and phospholipid-dependent pathways.[51] Since the activation of the sodium-proton-exchanger was demonstrated to be an early event in a series of rapid effects of aldosterone on HML electrolyte balance, the potential role of inositol-1,4,5-trisphosphate generation in short term effects of aldosterone was investigated.[52]

As determined by radioimmunoassay, the basal intracellular inositol-1,4,5-trisphosphate (IP_3) content of HML was 0.27 ± 0.02 pmol/10^6 cells. When HML were incubated with aldosterone (1 nM), intracellular IP_3 levels increased to a maximum of 0.58 ± 0.07 pmol/10^6 cells or $239 \pm 33\%$ of initial values after 60 sec ($p <0.05$), and declined to 0.37 ± 0.03 $p$mol/10^6 cells ($138 \pm 15\%$) after 5 min ($p <0.05$ vs. baseline control). Intracellular IP_3 levels were still significantly ($p <0.05$) above baseline values after 15 min (0.39 ± 0.05 pmol/10^6 cells; $145 \pm 15\%$). IP_3 levels after stimulation with concanavalin A (25 µg/ml) reached 0.53 ± 0.05 pmol/10^6 cells ($206 \pm 21\%$) after 60 sec ($p <0.05$), not significantly different from values 60 sec after aldosterone stimulation; however, after 5 and 15 min of incubation with concanavalin A, IP_3 levels were not different from baseline values.

Stimulation by increasing concentrations of aldosterone resulted in a dose-related response of IP_3 levels with half maximal effects (EC_{50}) of aldosterone at a concentration of ~0.1 nM.

Fludrocortisone was active at concentrations between 0.01 nM and 10 nM with an EC_{50} ~0.1 nM, thus similar to that of aldosterone. Hydrocortisone did not stimulate IP_3 generation up to a concentration of 10 nM. At 100 nM and 1000 nM, a minor stimulation of IP_3 was found; thus, the EC_{50} is estimated to be about 1 µM or above. Dexamethasone stimulates IP_3 generation starting at about 100 nM. Maximal IP_3 levels at 1000 nM dexamethasone were not different from levels after stimulation with 1 nM aldosterone. Incubation of the cells with 1 nM aldosterone plus 100 nM canrenone did not significantly inhibit the IP_3 response induced by aldosterone alone, and canrenone (100 nM) alone was inactive.

The main findings of this study on rapid aldosterone effects on IP_3 generation in HML are the following:

1. In HML, aldosterone significantly stimulates the generation of inositol-1,4,5-trisphosphate within 60 sec.
2. IP_3 stimulation was concentration dependent with an EC_{50} of ~0.1 nM.
3. Fludrocortisone had similar effects, while hydrocortisone and dexa methasone were partially active with an EC_{50} ~1 μM.
4. The classical mineralocorticoid inhibitor, canrenone, did not block aldosterone effects at 100-fold higher concentrations.

In terms of second messengers involved in rapid steroid effects, Sadler and Maller[53] have demonstrated that progesterone specifically inhibits adenylate cyclase in the Xenopus laevis oocyte plasma membrane and thus decreases cAMP levels. This effect appears to depend on a guanine nucleotide finding protein (G-protein). These findings were consistent with experiments by Maller and Krebs[54] showing that the level of cAMP-dependent protein kinase activity is the critical factor for the regulation of oocyte meiotic cell division. Functional studies in oocytes, spermatozoa, kidney cells, and neurons identified early changes in the intracellular calcium concentration as another possible intracellular second-messenger signal: Dufy et al.[68] reported an increased Ca^{2+}-dependent spiking activity in pituitary cells after estradiol application. Progesterone-induced Ca^{2+}-mobilization appears to be involved in meiotic maturation of the oocyte.[47] Similar effects on intracellular Ca^{2+}-concentration were found during acrosome reaction in spermatozoa.[48]

The linkage between the membrane receptor/aldosterone complex and stimulation of phospholipase C resulting in increased phosphoinositol breakdown and consequent IP_3 production is still unclear. Data by Orchinik et al.[43] suggested a possible involvement of guanine nucleotide binding proteins (G-proteins) in mediating agonist-receptor processes for steroid receptors of neuronal membranes. The binding of tritiated corticosterone to amphibian neuronal membranes was inhibited by nonhydrolyzable guanyl nucleotides, and the enhancement of inhibition by Mg^{2+} was consistent with the formation of a complex of steroid, receptor, and G-protein.

Steroid induced breakdown of phosphoinositides was first demonstrated by Steiner et al. in cultured rat VSMC.[49,55] The maximum rate of inositol-1,4,5-trisphosphate formation in VSMC treated by $LiCl_2$ was up to tenfold higher than basal levels at a concentration of 6,9 μmol/l hydrocortisone after 15 min. RU 486, a potent antiglucocorticoid, inhibited the hydrocortisone-induced response to only 50%.[55] Glucocorticoid actions different from those depending on transcriptional and translational processes, were tentatively postulated in that paper. For comparison with our data, it should be mentioned that a nonspecific inhibition of intracellular IP_3 breakdown was obtained by $LiCl_2$ in the previous paper, thus augmenting increases of IP_3 levels; given the similarity of those results compared with the findings obtained in HML, membrane receptors for mineralocorticoids may be occupied by glucocorticoids in concentrations higher than micromolar and inositol-1,4,5-trisphosphate formation in VSMC might be stimulated by 6.9 μM cortisol via the same non-genomic mechanisms as in HML.

The finding of an aldosterone selectivity of IP_3-stimulation is in agreement with both the functional and membrane binding data for aldosterone in HML, and again not compatible with the binding characteristics of the intracellular, cloned Type I mineralocorticoid receptor[3] which does not distinguish aldosterone from hydrocortisone. In addition, canrenone does not antagonize the aldosterone effect on IP_3-stimulation, again consistent with data for aldosterone effect on Na^+/H^+-stimulation and aldosterone membrane binding.

Activation of lymphocytes by lectins, such as concanavalin A or phytohaemagglutinin, involves the stimulation of the phospholipase C turnover and concomitant increases of intracellular calcium due to influx of extracellular calcium.[56,57] In the case of the aldosterone effect in HML, rapid increases of free intracellular calcium by aldosterone are not seen, indicating the absence of a detectable, instant stimulation of transmembrane calcium fluxes.

In conclusion, aldosterone selectively increases intracellular inositol-1,4,5-trisphosphate levels in HML with an apparent K_d of about 0.1 nM, and hydrocortisone and dexamethasone are active only at much higher concentrations. These findings strongly support the hypothesis of a novel rapid pathway for aldosterone action whereby mineralocorticoids may be involved in short-term cardiovascular regulation via the phospholipid second messenger system.

b. Free Intracellular Calcium

To further investigate the post-receptor mechanisms of the mineralocorticoid action, the effects of aldosterone on the free intracellular calcium as a possible messenger for cardiovascular effects were studied *in vitro* by the fluorescent dyes Quin2 and Fura2 in HML.[28] If determined by Quin2, incubation of HML with 2.8 nM aldosterone for 1 h at 37°C significantly increased free intracellular calcium $[Ca]^{2+}_i$ to 139 ± 39 nM compared with 118 ± 27 nM obtained without aldosterone added (p <0.05). Basal $[Ca]^{2+}_i$ in freshly isolated HML was 54 ± 15 nM (Fura2). After incubation without aldosterone $[Ca]^{2+}_i$ significantly fell to 50 ± 13 nM (p <0.05). After incubation with 1.4 nM aldosterone $[Ca]^{2+}_i$ remained constant at 57 ± 14 nM (*p* <0.05 compared with value after incubation without aldosterone). 140 nM canrenoate or 700 nM canrenone did not antagonize the action of aldosterone while 1 μM *N*-ethylisopropylamiloride (EIPA) completely blocked the effect.

Half maximal effects of aldosterone were seen at concentrations between 0.1 and 1 nM. The effect of 14 nM aldosterone was not antagonized by the presence of actinomycin D (50 μg/l) or cycloheximide (20 μg/ml).

The main findings of this study are:

- Aldosterone prevents the decrease of free intracellular calcium in human mononuclear leukocytes, which is observed after 1 h of incubation at 37°C without aldosterone.
- The effect of aldosterone is antagonized by *N*-ethyl-isopropylamiloride, but not by the mineralocorticoid antagonists canrenoate and canrenone, and by the inhibitors of RNA or protein synthesis, actinomycin D, and cycloheximide.

The extent of changes in $[Ca]^{2+}_i$ by aldosterone (14 to 18%) is small if compared with other agonists which elevate $[Ca]^{2+}_i$ by more than 100%, such as angiotensin II.[58] However, small variations in intracellular electrolytes may still have a physiological significance for the tonic regulation of various electrolyte-dependent processes. As an example for a tonic regulation of cell function by small changes of intracellular electrolytes, the effects of digitalis glycosides on the intracellular sodium is mentioned here which does not exceed 10 to 20% at a maximum inotropic response of myocardial tissue to ouabain.[59]

These findings support the assumption that mineralocorticoids concordantly increase the concentrations of both the intracellular sodium[26] and the free intracellular calcium. Among others, one possible mechanism for this parallel change of both ions would be the involvement of the Na^+-Ca^{2+}-exchanger. This hypothesis is supported by the lack of an aldosterone effect on $[Ca]^{2+}_i$ in sodium-free buffer, in which sodium cannot enter the cells and activate the Na^+-Ca^{2+}-exchanger. However, conflicting results are reported about the existence of this transporter in HML.[57,60,61] In one study only, a Na^+-Ca^{2+}-exchange was demonstrated in HML membrane vesicles.[31]

The effect of aldosterone on $[Ca]^{2+}_i$ shares similarities with the action of aldosterone on intracellular sodium, potassium, and cell water: the half-maximal effect on these parameters was also found between 0.14 and 1.4 n*M*. However, these effects were at least partially antagonized by canrenone, while with regard to $[Ca]^{2+}_i$, the mineralocorticoid antagonists canrenone and canrenoate were ineffective.

This discrepancy between findings on effects of aldosterone on $[Ca]^{2+}_i$ and sodium, potassium, and cell volume could possibly reflect differences in early and late effects of mineralocorticoids: the first step in the electrolyte effects of aldosterone in HML involves the non-genomic activation of the sodium-proton exchanger, which also was insensitive to canrenone. The subsequent steps in the electrolyte effects of aldosterone (activation or recruitment/production of new molecules of the Na-K-ATPase with a concordant increase of potassium and cell water) seem to be at least partially canrenone-sensitive and, thus, may in part correspond to the known, genomic effects of mineralocorticoids in classical target organs, e.g., renal tubules.

The insensitivity of the increase of $[Ca]^{2+}_i$ by aldosterone to cycloheximide and actinomycin D also supports the assumption that this effect is non-genomic. The sodium-proton exchanger is inhibited by amiloride or N-ethyl-isopropylamiloride at the concentration employed here (1 μM).[62,63] That would explain the antagonist effect of N-ethyl-isopropylamiloride on the aldosterone-induced increase of $[Ca]^{2+}_i$ as a consequence of the inhibition of the sodium-proton exchanger.

These data demonstrate that the moderate increase of free intracellular calcium in HML is related more closely to early events of aldosterone effects on electrolyte transport such as an activation of the sodium-proton exchanger, while 1 h effects on HML sodium, potassium, and cell volume include canrenone-sensitive, presumably genomic effects which may augment initial rapid changes in electrolyte transport. Latter effects may be experimentally undetectable with regard to changes in gross electrolyte content as a net result of multiple processes, because of their small extent and/or counterregulatory processes which antagonize the changes of strictly controlled parameters with vital impact on cell function, such as electrolyte concentrations or cell volume.

IV. A NEW DUAL MODEL FOR ALDOSTERONE ACTION

Evidences for rapid steroid action, which cannot be mediated by traditional, genomic mechanisms for various reasons such as time course, resistance to inhibitors of transcription, and protein synthesis, have been occasionally reported in the literature for the last two decades. However, as mentioned in the introduction section, these data have been neglected in the scientific discussion since they were incompatible with the prevalent dogma of the theory of steroid action. Purists still argued against the assumption of a non-genomic pathway for steroid action when rapid effects started within 15 min. Their main motive was that very early genomic events may be seen after few minutes, for example, with regard to MMTV LTR transcription under the influence of steroids (see above). These arguments appear to be overstressed since the initiation of an effector process at the membrane level requires a long chain of preceding events (in its shortest case, binding to cytosolic receptors, binding of the complexes to nuclear DNA, production of m-RNA, translation to proteins and, finally, subsequent activation of electrolyte transport systems; see introduction). No example for such a complete composite response mechanism for steroids can found in the literature for a range of 15 to 30 min. In the case of mineralocorticoids, genomic responses typically start 2 to 4 h after the application of aldosterone, and reach a plateau after 24 or more hours. This range of time for aldosterone action applies to effects on transepithelial electrolyte transport, e.g., aldosterone effects on short circuit current, transepithelial sodium transport in renal or bladder epithelia.[64]

To our knowledge, a dual model for mineralocorticoid action was first hypothesized tentatively by Moura and Worcel.[17] Both the primary increase in sodium permeability and the early, secondary stimulation of the Na-K-ATPase by aldosterone in rat tail smooth muscle cells, were insensitive to actinomycin D and, therefore, apparently represented a non-genomic response; after >1 h, an additional increase in sodium efflux is observed which is sensitive to actinomycin D and thus likely reflects a genomic response. Its molecular correlate is the *de novo* synthesis of Na-K-ATPase molecules. The initial increase of active sodium efflux is thought to represent the secondary stimulation of the pre-existing molecules of the Na-K-ATPase by the primary, non-genomically transmitted increase in sodium influx. Unfortunately, the authors did not continue to study these rapid aldosterone effects in rat smooth muscle cells and elucidate the underlying mechanisms. Thus, no widely accepted support for the theory of non-genomic steroid action has come from these very mature and valuable observations. In that pivotal paper, a relatively indiscriminatory compilation of four possible explanations for the rapid aldosterone effects, including a genomic and a non-genomic mode of action, is given. This left the question of the non-genomic nature of these effects open for discussion. Similar problems are immanent to the initial findings by Oberleithner et al.[39] who showed changes in intracellular pH 15 to 20 min after application of aldosterone. However, one should not forget that intracellular pH is buffered and changes induced by any effector, e.g. the sodium-proton exchanger can be seen only after the buffer capacity is overrun. This may require additional time and explain a latency of that extent.

In HML and VSMC, the activation of the sodium-proton exchanger, and in HML, the stimulation of IP_3-generation are very rapid. Effects are seen within 1 to 4 min and, thus, are due to a non-genomic steroid effect with certainty, as discussed above. Thus, these findings are the first to convincingly show a non-genomic response to aldosterone. This response has been intensively characterized and, by means of various studies on different effector systems in different cell types, found to be transmitted by a receptor/effector mechanism which is completely distinct from the traditional genomic pathway of steroid action.

Though not definitively shown, but suggested for HML by a differential effect of canrenone in terms of effects after minutes or hours in HML, mineralocorticoids certainly also exert genomic responses, as extensively studied in the classical target tissue, the distal kidney tubule[9] and toad bladder.[65]

From these traditional genomic effects and the recent findings on rapid, non-genomic effects of aldosterone, the following model of mineralocorticoid action is derived (Figure 12):

1. The primary steps involve steroid binding to membrane receptors which have been identified in HML for the first time. These receptors trigger changes of electrolyte transport systems in the membrane. This response is fast and starts within 1 to 2 min. Secondary to this step, the immediate activation of pre-existing electrolyte transport systems, e.g., the Na-K-ATPase, is induced reactively by changes in electrolyte concentrations. This effect is also fast and starts, without a major delay, within minutes.

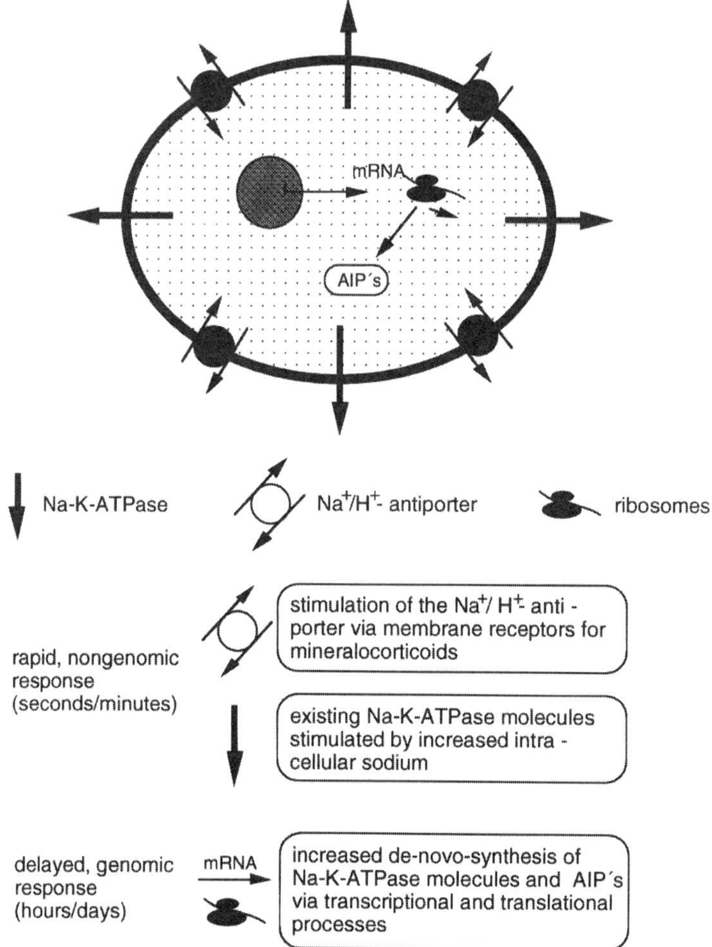

Figure 12. The primary step of mineralocorticoid action (increased sodium influx into the cell by the sodium-proton-exchanger) and the subsequent cascade of effects on sodium fluxes and Na-K-ATPase production involving a membrane receptor for mineralocorticoids are shown. These steps are related to the early and late responses of electrolyte fluxes. AIP's: aldosterone induced proteins. (Adapted from Wehling, M., Christ, M., Gerzer, R., *Trends Pharmacol. Sci.,* 14, 1, 1993. With permission.)

2. A completely different response involves genomic mechanisms and starts after a latency of one or more hours. This response includes the production of AIPs (see above) and new Na-K-ATPase molecules. The first steps of late membrane effects, secondary to genomic events involving protein synthesis, include the increase of conductive sodium transport which was described as early as 1963[65] for renal cells. As HML are commonly considered to bear only small, if any significant numbers of

sodium "channels" this transport system is not expected to play a major role for aldosterone effects in HML.

It still has to be determined how the non-genomic and genomic effects of mineralocorticoids and other steroids are interrelated. For example, the question should be addressed if the synthesis of new Na-K-ATPase molecules under the influence of mineralocorticoids at least in part represents an adaptive up-regulation of protein synthesis or not. This adaptation would reflect a response to a persistent cellular sodium load caused by the activated sodium-proton exchanger. In HML, genomic responses appear to augment initial non-genomic effects of aldosterone after a latency in the range of 1 h. This genomic augmentation, thus, might be responsible for experimentally detectable net changes of intracellular electrolytes and cell volume. Functionally, this would mean an enhancement of a tonic, constant activation of the non-genomic effector mechanism while rapid "clonic" changes of the fast response system would not be augmented. Latter changes, thus, could be involved in a rapid controlling mechanism in various tissues, e.g., the cardiovascular system.

The model presented here for mineralocorticoids shares large similarities with the theory developed almost simultaneously for the action of steroids at the membrane level in the brain: it is not necessary to reiterate the essentials from Chapter 4 here.

As mentioned above, the actions of 5α-pregnan-3α,21-diol-20-one and 5α-pregnan-3α-OH-20-one on the $GABA_A$-receptor complex, the effects of pregnanolone on LHRH release, of progesterone on dopamine release, certain neural effects by the local application of steroids, and the action of progesterone on the oocyte maturation and spermatozoan acrosome reaction are likely candidates to occur *in vivo* if physiological concentrations of the steroidal agonists are considered. This implies that for other steroid-membrane interactions, supraphysiological concentrations are required and, therefore, their physiological meaning still has to be determined (see Tables 2 and 3). For these rapid steroid effects, a similar model of steroid action might apply, though the genomic components in these responses have not been separated from the non-genomic steroid action as clearly as in the case of the action of aldosterone in various tissues. Therefore, it is possible that in other systems only part one (non-genomic action) or two (genomic action) of the two-step model of steroid action is realized, and further studies on this issue are eagerly awaited.

V. CLINICAL IMPLICATIONS

With regard to sodium and potassium content, the HML model has been applied to various clinical situations in patients with disturbances of sodium and water balances. At the level of net changes of these cellular electrolytes after 1 h, abnormalities in the response to aldosterone have been found in patients with pseudohypoaldosteronism, primary and secondary aldosteronism,

TABLE 2
Examples for Non-Genomic Steroid Effects

Steroid	Effect	K_d	Ref.
Estradiol	Endometrial cells: effects on Ca^{2+}-influx	~1 nM	66
Estradiol	Pre-optical-septal neuron (iontophoresis)	?	67
Estradiol	Spiking-activity of pituitary cells	~1 nM	68
Progesterone	$[Ca^{2+}]_i$ in oocytes	~3 μM	47
Aldosterone	^{22}Na-efflux from rat aorta VSMC	?	17
Progesterone	Acetylcholine-release	~0.3 μM	69
Aldosterone	Intracellular pH in kidney cells	300 nM	39
Pregnalonolone	LHRH-release	0.03 nM	70
Progesterone	Acrosome reaction in sperms	~0.1 μM	48
Vitamin D_3	Rapid increase of cGMP in human fibroblasts	~0.1 nM	71
Aldosterone	Na^+/H^+-antiport in human lymphocytes	0.04 nM	34
Corticosterone	Modification of sexual behavior in male frogs	11 μM	43
Aldosterone	Rapid increase of $[Ca^{2+}]_i$ in renal cells	?	72
17-β-Estradiol	Intracellular $[Ca^{2+}]_i$ in granulosa cells	~0.1 nM	73

TABLE 3
Steroid Binding to Plasma Membranes

Steroid and Binding Substrate	K_d	Ref.
Cortisol: liver	~1.5 nM	74
17-β-estradiol: liver and endometrium	~1 nM	66
2-OH-estradiol: pituitary cells	0.4 nM	75
R5020: Xenopus laevis oocytes	1.2 μM	76
Estradiol: pituitary	0.04 nM	77
Dexamethasone: liver	400 nM	78
Progesterone: neural membranes	~30 nM	79
Corticosterone: membranes from frog brain	~0.5 nM	43
R5020: rat liver	33 nM	80
Aldosterone: human lymphocytes	0.1 nM	41
Aldosterone: rat kidney	3–14 nM	44
Aldosterone: rat kidney	100 nM	45

and essential hypertension. They were compared with findings on classical Type I mineralocorticoid receptors by Armanini et al.[83,84,88]

In seven patients with pseudohypoaldosteronism, the effects of aldosterone on intracellular sodium and potassium were studied and compared with normal controls, in whom aldosterone prevents the loss of sodium and potassium *in vitro*. In the patients, intracellular sodium and potassium decreased normally in the absence of aldosterone. With aldosterone added to the incubation medium, intracellular sodium and potassium were not different from values obtained without aldosterone. Thus, incubation with aldosterone did not affect intracellular sodium and potassium, as would be seen in normals. Baseline values of sodium and potassium before the incubation were within the normal range.

From these findings it is concluded that, after the critical period of the disease during the first months of life, intracellular sodium and potassium may be maintained at normal levels by mechanisms unrelated to the action of mineralocorticoids. Though not yet known in detail, these mechanisms are quickly reversible, since a significant wash-out of the sodium increasing effect is observed within 1 h of incubation with or without aldosterone. These mechanisms may include effects of salt supplementation, or a resetting of the sensitivity of central nervous centers which regulate electrolyte and water balance directly or by endogenous factors different from aldosterone.[81] Additionally, the families of seven patients with pseudohypoaldosteronism (index cases) were studied. In the first family studied, two siblings were affected by the disease and had a reduced number of mineralocorticoid Type I receptors on HML. Intracellular sodium and potassium in HML from these patients did not show a response to 1.4 nM aldosterone. The parents, who were first cousins, had no history of disease and normal receptor data, but in the mother, the response of HML electrolytes to aldosterone was abnormal. In the second family, the mother of a child with pseudohypoaldosteronism, the mother's sister, and her son had low numbers of Type I receptors. Only the aunt of the index case had an uncertain history of the disease. The mineralocorticoid effector mechanism was abnormal in both children and both mothers studied. In a third family, the effector defect was present only in HML of the father. In three other families, the abnormality of the effector mechanism was detected in HML of the patients' mothers.[82]

These data suggest an interpretation different from that derived from the original data on HML Type I receptors by Armanini et al.[83,84] who suggested two different modes of inheritance: an autosomal-dominant and an autosomal-recessive trait. Since the defect at the level of the mineralocorticoid effector mechanism was present in one parent of all patients studied, the mode of inheritance appears to be dominant.

This assumption is supported by earlier reports of Roy,[85] Limal et al.,[86] and by Hanukoglu et al.[87] Roy[85] found high aldosterone levels in the healthy father of four affected children. Limal et al.[86] described a family with affected members in three consecutive generations. Hanukoglu et al.[87] examined the healthy relatives of an affected girl and found high plasma aldosterone and renin levels in two brothers, the patient's mother, and her maternal grandmother. However, determination of plasma aldosterone and renin levels may not be sensitive enough to perform conclusive genetic studies using plasma aldosterone and renin levels as markers.

In the first family studied, both parents had normal type-I receptors in HML, but the patient's mother showed the abnormal electrolyte response in HML to aldosterone. This finding supports the assumption that, at least in this family, the primary defect involves the aldosterone effects at the effector level, rather than at the receptor level, if receptor numbers are considered.

In the second kindred studied, the absence or decreased number of Type I receptors, which was also found in relatives free of symptoms, could partly reflect a compensatory "down-regulation" in response to the elevated aldosterone levels. This mechanism has been shown to apply to patients with high aldosterone levels due to Conn's disease or other causes of aldosteronism.[88] This explanation would assume that the absent or low number of Type I receptors represents a secondary phenomenon in consequence to chronic hyperaldosteronemia in these patients.

It is notable, that for the Type I receptor, as well as for the aldosterone effector abnormalities, no close relation to the clinical symptoms or a history of the disease was found. The first degree cousin of the index case of the second family was not affected by pseudohypoaldosteronism, but had an abnormal number of Type I receptors and an abnormal electrolyte response. Additionally, plasma aldosterone and renin levels were not abnormal in our healthy gene carriers and, therefore, do not appear to be a sensitive tool for inheritance studies of this disease. Compensatory mechanisms have to be postulated which are responsible for an almost normal sodium balance, plasma aldosterone, and renin levels in these individuals, but are not yet known in detail. Such compensatory mechanisms, as well as a variable gene expression, may result in the wide range of clinical presentations which further support a dominant trait.

These results were important to validate the HML as a useful cell model to study extrarenal mineralocorticoid effector mechanisms. Thus, a "black and white" experiment of nature served as a positive control for this model.

The discrepancy between findings about HML Type I receptors and aldosterone effects in patients with pseudohypoaldosteronism may indicate that the molecular base of the disease could include abnormalities of the aldosterone membrane receptors and rapid effects of aldosterone. As shown above, these rapid effects could be responsible for the increased influx of sodium into cells, most likely via an activation of the sodium-proton exchanger. So far, 1-h effects only have been studied in these patients, which may include genomic responses as well as non-genomic mechanisms. Studies focusing on aldosterone membrane receptors and rapid effects are presently being performed.

In patients with primary and secondary aldosteronism, an abnormal response of HML sodium and potassium was found. There was no decrease of HML sodium and potassium during incubation without aldosterone, and aldosterone did not affect these electrolytes if present in the medium. It was hypothesized that this abnormal response may reflect the cellular correlate of the "escape" phenomenon in patients with aldosteronism. This term was used to describe the renal resistance to excess mineralocorticoids in chronic states of aldosteronism which blunts the hypokalemic and antinatriuretic effects of mineralocorticoids.[89]

The effect of aldosterone on HML sodium and potassium was investigated in 13 patients with essential hypertension. In only four patients, sodium in HML without incubation was elevated compared with the range for normal persons. A decrease of intracellular sodium or potassium occurred during incubation without aldosterone. In contrast to effects in HML from normals, the addition of 1.4 n*M* aldosterone did not prevent this loss of electrolytes.

Plasma renin activity and aldosterone were not correlated with the electrolyte response and were within the normal limits. The number of "classical" Type I mineralocorticoid receptors/cell were within, or close to, the normal range.[90]

These findings indicate an independence of intracellular electrolytes in HML of patients with essential hypertension from the mineralocorticoid receptor-mediated effector mechanism by which intracellular sodium and potassium are elevated *in vivo* and *in vitro*. The most attractive explanation for this would be a counter-regulatory impairment of the mineralocorticoid effector mechanism which is normally responsible for an elevation of intracellular electrolytes. Though being speculative, this impairment would blunt the increased sodium influx and intracellular sodium concentration due to a genuine membrane defect in essential hypertension. Though discussed controversially, most known membrane defects in essential hypertension would result in an elevated intracellular sodium. It is notable that increased transmembranal sodium fluxes and intracellular sodium in lymphocytes of essential hypertensives have been reported earlier by Edmondson et al.[91] and Ambrosioni et al.[92] Latter authors found clearly distinct values for normal persons and hypertensives without overlap. In contrast, our results show that even if a significant elevation of intracellular sodium would be detected in a larger group of patients an essential overlap is present. The gap between the four markedly elevated intracellular sodium values and the normal values may suggest an inhomogeneous distribution of elevated and normal values. The independence of the impairment of the mineralocorticoid effect from the number of Type I mineralocorticoid receptors/cell shows that this defect is not just due to a "down-regulation" of mineralocortcoid receptors as observed in primary aldosteronism.[88] Plasma aldosterone and renin activity appeared not to be involved, thus indicating that the impairment of the mineralocorticoid effector mechanism is due to an unknown or undetermined factor, presumably at the subcellular level. This could include an abnormality of the novel membrane receptor for aldosterone and/or the related effector mechanism. In this context, the activity of the sodium-proton exchanger in HML from essential hypertensives was studied. A stimulation of the sodium-proton exchanger has been shown earlier in platelets of hypertensive man, and lymphocytes of spontaneous hypertensive rats.[93,94] HML were investigated in 12 patients with essential hypertension with regard to the activity of the sodium-proton exchanger and HML volume. Compared with matched normotensives, the cell volume of HML in a physiological buffer was significantly increased in essential hypertension (p <0.05). The amiloride-inhibitable rate of cell swelling in isotonic sodium propionate was also increased in HML from hypertensives. 400 nM amiloride abolished the difference of cell volume within 1 min.[95] This activation of the sodium-proton exchanger was independent from the effect of aldosterone.

These data show a functional swelling of HML in essential hypertension, probably due to an activation of the sodium-proton exchanger. If representative also for VSMC, these findings could explain hypertensive vessel wall hypertrophy in part as functional cell swelling. At least for the abnormality of 1-h effects of aldosterone in essential hypertensives, the rapid aldosterone effect does not appear to be involved.

The investigations of the easily accessible HML model in clinical situations may underline its applicability to pathological states of the sodium and water balances and its use in the study of the involvement of mineralocorticoids in these diseases. However, the interpretation of the results has to take into account that their major part was obtained at the level of 1-h effects of aldosterone on HML sodium and potassium. Since a clear distinction of genomic and non-genomic aldosterone effects is not possible at this time, more data are required to identify the involvement of the non-genomic mechanism in these intermediate effects.

The findings on aldosterone summarized here add another steroidal effect mediated by membrane receptors to the array mentioned above. The possible physiological significance of a non-genomic aldosterone action is underlined by the low K_d for both the rapid *in vitro* effects and membrane binding of aldosterone, which are in agreement with the physiological concentration of free aldosterone *in vivo* (~0.1 nM in humans, ~0.2 nM in rats). The aldosterone-selectivity of the rapid *in vitro* effects and membrane binding of aldosterone should be pointed out here since it offers a new explanation for the apparent clinical evidence for differential effects of glucocorticoids vs. mineralocorticoids. The concentrations necessary for both the effects and binding of aldosterone in HML and VSMC differ from that of cortisol by 3 to 5 orders of magnitude, thus attaching the aldosterone-selectivity to the level of the membrane receptor. This would complement the recently acknowledged role of the renal 11-β-hydroxysteroid dehydrogenase[96] as a mechanism ensuring mineralocorticoid specificity. The significance and relation of both mechanisms to confer steroid-specificity has to be re-evaluated in the light of the findings summarized here.

The rapid, non-genomic mechanism of aldosterone action could identify this steroid as a hormone with cardiovascular relevance. From a teleological point of view, this mechanism could explain the rapid postural change of aldosterone plasma concentrations as an acute regulatory response to maintain the circulatory homeostasis by aldosterone as a pressure hormone. This function could be hypothesized from the impact of aldosterone on VSMC sodium metabolism leading to an increased intracellular sodium concentration and, via the sodium-calcium exchange, to increased vascular tone. In this context, old clinical observations should be re-called: the management of Addison's crisis cannot be accomplished by electrolyte- and water replacement of renal losses, but has to include the hormonal substitution. This most likely reflects the cardiovascular effects aside from renal actions which cannot be substituted by electrolyte and water replenishment alone.

Further investigations on the function of aldosterone membrane receptors will have to focus on their possible involvement in human physiology and pathophysiology with regard to electrolyte transport, cardiovascular regulation and, even, immune response. Major results have been obtained in lymphocytes with their physiological significance being completely unknown so far in these cells.

The findings reported here could lead to the development of a membrane receptor antagonist for mineralocorticoids. Antagonists of aldosterone at the level of the cell membrane may be the base of new strategies in the treatment of hypertension and heart failure in man.

ACKNOWLEDGMENTS

Part of the author's studies reported here were supported by the "Wilhelm-Sander-Stiftung" (88.015.2/3) and the "Deutsche Forschungsgemeinschaft" (We 1184/4-1/2).

REFERENCES

1. **Langford, H. G., Snavely, J. R.,** Effect of DOCA on development of renoprival hypertension, *Am. J. Physiol.,* 196, 449, 1959.
2. **Armanini, D., Strasser, T., Weber, P. C.,** Characterization of aldosterone binding sites in circulating human mononuclear leukocytes, *Am. J. Physiol.,* 248, E388, 1985.
3. **Arriza, L. A., Weinberger, C., Cerelli, G., Glaser, T. M., Handelin, B. L., Housman, D. E., Evans, R. M.,** Cloning of human mineralocorticoid receptor complementary DNA: structural and functional kinship with the glucocorticoid receptor, *Science,* 23, 268, 1987.
4. **Vander, A. J., Malvin, R. L., Wilde, W. S., Lapides, J., Sullivan, L. P., McMurray, V. M.,** Effects of adrenalectomy and aldosterone on proximal and distal tubular sodium reabsorption, *Proc. Soc. Exp. Biol. Med.,* 99, 323, 1958.
5. **Edelman, I. S., Fimognari, G. M.,** Modes of hormone action: on the biochemical mechanism of action of aldosterone, *Recent Prog. Horm. Res.,* 24, 1, 1968.
6. **Myers, J. H., Bohr, D.,** Mechanisms responsible for the pressure elevation in sodium dependent mineralocorticoid hypertension, in *Endocrinol. Hypertension,* Mantero, F., Biglieri, E. G., and Edwards, C. R. W., Eds., Raven Press, New York, 1985, 131.
7. **Bohr, D. F., Harris, A. L., Guthe, C., Webb, R. C.,** Hypertension: multiple membrane malfunctions, in *Topics in Pathophysiology of Hypertension,* Villareal, H. and Sambhi, M. P., Eds., Martinus Nihnhoff, Amsterdam, 1984, 101.
8. **Garty, H.,** Mechanism of aldosterone action in tight epithelia, *J. Membr. Biol.,* 90, 193, 1986.
9. **Marver, D., Kokko, J. P.,** Renal target cells and the mechanism of action of aldosterone, *Miner. Electrolyte Metab.,* 9, 1, 1983.
10. **Miller, A. W., Bohr, D. F., Schork, M. A., Terris, J. M.,** Hemodynamic responses to DOCA in young pigs, *Hypertension,* 1, 591, 1979.
11. **Conway, F. J., Hatton, R.,** Development of deoxycorticosterone acetate hypertension in the dog, *Circ. Res.,* 43, 82, 1978.
12. **Pan, Y., Young, D. B.,** Experimental aldosterone hypertension in the dog, *Hypertension,* 4, 279, 1982.
13. **Jones, A. W.,** Kinetics of active sodium transport in aortas from control and deoxycorticosterone hypertensive rats, *Hypertension,* 3, 631, 1981.
14. **Jones, A. W., Hart, R. G.,** Altered ion transport in aortic smooth muscle during deoxycorticosterone acetate hypertension in rat, *Circ. Res.,* 37, 333, 1975.

15. **Friedman, S. M., Friedman, C. L.,** Cell permeability, sodium transport, and the hypertensive process in the rat, *Circ. Res.,* 39, 441, 1976.

16. **Kornel, L., Kanamarlapudi, N., Ramsey, C., Travers, T., Kamath, S., Taff, D. J., Patel, N., Packer, W., Raynor, W. J.,** Arterial steroid receptors and their putative role in the mechanism of hypertension, *J. Steroid Biochem.,* 19, 333, 1983.

17. **Moura, A. M., Worcel, M.,** Direct action of aldosterone on transmembrane Na$^+$ efflux from arterial smooth muscle, *Hypertension,* 6, 425, 1984.

18. **Christ, M., Douwes, K., Theisen, K., Wehling, M.,** Rapid aldosterone action on the Na$^+$/H$^+$-antiport of vascular smooth muscle cells: a new pathway of steroid action, *J. Am. Coll. Cardiol.,* 21, 162A, 1993.

19. **Frelin, C., Vigne, P., Lazdunski, M.,** The Na$^+$/H$^+$ exchange system in vascular smooth muscle cells, *Adv. Nephrol.,* 19, 17, 1990.

20. **Berk, B. C., Alexander, R. W.,** Vasoactive effects of growth factors, *Biochem. Pharmacol.,* 38, 219, 1989.

21. **Berk, B. C., Taubman, M. B. , Griendling, K. K., Cragoe, E. J., Fenton, W., Brock, T. A.,** Thrombin-stimulated events in cultured vascular smooth-muscle cells, *Biochem. J.,* 274, 799, 1991.

22. **Vallega, G. A., Canessa, M. L., Berk, B. C., Brock, T. A., Alexander, R. W.,** Vascular smooth muscle Na$^+$-H$^+$ exchanger kinetics and its activation by angiotensin II, *Am. J. Physiol.,* 254, C751, 1988.

23. **Berk, B. C., Vallega, G., Griendling, K. K., Gordon, J. B., Cragoe, E. J., Canessa, M., Alexander, R. W.,** Effects of glucocorticoids on Na$^+$/H$^+$ exchange and growth in cultured vascular smooth muscle cells, *J. Cell Physiol.,* 137, 391, 1988.

24. **Al-Dujaili, E. A. S., Edwards, C. R. W.,** The development and application of direct radioimmunoassay for plasma aldosterone using 131-I-labelled ligand-comparison of three methods, *J. Clin. Endocrinol. Metab.,* 46, 105, 1978.

25. **Menachery, A., Braley, L. M., Kifor, I., Gleason, R., Williams, G. H.,** Dissociation in plasma renin and adrenal ANG II and aldosterone responses to sodium restriction in rats, *Am. J. Physiol.,* 261, E487, 1991.

26. **Wehling, M., Armanini, D., Strasser, T., Weber, P. C.,** Effect of aldosterone on the sodium and potassium concentrations in human mononuclear leukocytes, *Am. J. Physiol.,* 252, E505, 1987.

27. **Wehling M., Kuhls, S., Armanini, D.,** Volume regulation of human lymphocytes by aldosterone in isotonic media, *Am. J. Physiol.,* 257, E170, 1989.

28. **Wehling M., Käsmayr, J., Theisen, K.,** Aldosterone influences free intracellular calcium in human mononuclear leukocytes *in vitro, Cell Calcium,* 11, 565, 1990.

29. **Grinstein, S., Clarke, C. A., Rothstein, A.,** Volume-induced increase of anion permeability in human lymphocytes, *J. Gen. Physiol.,* 82, 619, 1983.

30. **Blaustein, M. P.,** Sodium ions, calcium ions, blood pressure regulation and hypertension: a reassessment and a hypothesis, *Am. J. Physiol.,* 232, C165, 1977.

31. **Ueda, T.,** Na$^+$ - Ca^{++} exchange activity in rabbit lymphocytes plasma membranes, *Biochim. Biophys. Acta,* 734, 342, 1983.

32. **Grinstein, S., Goetz, J. D., Furuya, W., Rothstein, A., Gelfand, E. W.,** Amiloride-sensitive Na$^+$/H$^+$ exchange in platelets and leukocytes: detection by electronic cell sizing, *Am. J. Physiol.,* 247, C293, 1984.

33. **Wehling, M., Käsmayr, J., Theisen, K.,** Fast effects of aldosterone on electrolytes in human lymphocytes are mediated by the sodium-proton-exchanger of the cell membrane, *Biochem. Biophys. Res. Commun.,* 164, 961, 1989.

34. **Wehling, M., Käsmayr, J., Theisen, K.,** Rapid effects of mineralocorticoids on sodium-proton-exchanger: genomic or nongenomic pathway?, *Am. J. Physiol.,* 260, E719, 1991.

35. **Minuth, W. W., Steckelings, U., Gross, P.,** Complex physiological and biochemical action of aldosterone in toad urinary bladder and mammalian renal collecting duct cells, *Renal Physiol.,* 10, 297, 1987.

36. **Cato, A. C. B., Weinmann, J.,** Mineralocorticoid regulation of transcription of transfected mouse mammary tumor virus DNA in cultured kidney cells, *J. Cell Biol.*, 106, 2119, 1988.

37. **Groner, B., Hynes, N. E., Rahmsdorf, J., Ponta, H.,** Transcription initiation of transfected mouse mammary tumor virus LTR DNA is regulated by glucocorticoid hormones, *Nucl. Acids Res.*, 11, 4713, 1983.

38. **Geering, K., Michel, C., Gaeggeler, H. P., Rossier, B. C.,** Receptor occupancy vs. induction of Na^+-K^+-ATPase and Na^+ transport by aldosterone, *Am. J. Physiol.*, 248, C102, 1985.

39. **Oberleithner, H., Weigt, M., Westphale, H.-J., Wang, W.,** Aldosterone activates Na^+/H^+ exchange and raises cytoplasmic pH in target cells of the amphibian kidney, *Proc. Natl. Acad. Sci. U.S.A.*, 84, 1464, 1987.

40. **Wehling, M., Christ, M., Theisen, K.,** Evidence for high affinity mineralocorticoid receptors in plasma membrane rich preparations from human lymphocytes, *Biochem. Biophys. Res. Commun.*, 181, 1306, 1991.

41. **Wehling, M., Christ, M., Theisen, K.,** Membrane receptors for aldosterone: a novel pathway for mineralocorticoid action, *Am. J. Physiol.*, 263, E974, 1992.

42. **Schmidt-Ullrich, R., Wallach, D. F. H., Davis, F. D. G. II,** Membranes of normal hamster lymphocytes and lymphoid cells neoplastically transformed by Simian Virus 40. High yield purification of plasma membrane fragments, *J. Natl. Cancer Inst.*, 57, 1107, 1976.

43. **Orchinik, M., Murray, T. F., Moore, F. L.,** A corticosteroid receptor in neuronal membranes, *Science*, 251, 1848, 1991.

44. **Ozegovic, B., Dobrovic-Jenik, D., Milkovic, S.,** Solubilization of rat kidney plasma membrane proteins associated with ^{3}H-aldosterone, *Exp. Clin. Endocrinol.*, 92, 194, 1988.

45. **Forte, L. R.,** Effect of mineralocorticoid agonists and antagonists on binding of ^{3}H-aldosterone to adrenalectomized rat kidney plasma membranes, *Life Sci.*, 11, 461, 1972.

46. **Wehling, M., Eisen, C., Aktas, J., Christ, M., Theisen, K.,** Photoaffinity labeling of plasma membrane receptors for aldosterone from human mononuclear leucocytes, *Biochem. Biophys. Res. Commun.*, 189, 1424, 1992.

47. **Wassermann, W. J., Pinto, L. H., O´Connor, C. M., Smith, L. D.,** Progesterone induces a rapid increase in $[Ca^{2+}]$in of Xenopus laevis oocytes, *Proc. Natl. Acad. Sci. U.S.A.*, 77, 1534, 1980.

48. **Blackmore, P. F., Beebe, S. J., Danforth, D. R., Alexander, N.,** Progesterone and 17α-hydroxyprogesterone. Novel stimulators of calcium influx in human sperm, *J. Biol. Chem.*, 265, 1376, 1990.

49. **Steiner, A., Vogt, E., Locher, R., Vetter, W.,** Stimulation of the phosphoinositide signaling system as a possible mechanism for glucocorticoid action in blood pressure control, *J. Hypertension*, 6, S366, 1988.

50. **Sato, A., Suzuki, H., Iwaita, Y., Nakazato, Y., Kato, H., Saruta, T.,** Potentiation of inositol trisphosphate production by dexamethasone, *Hypertension*, 19, 109, 1992.

51. **Mahnensmith, R. L., Aronson, P. S.,** The plasma membrane sodium-hydrogen exchanger and its role in physiological and pathophysiological processes, *Circ. Res.*, 56, 773, 1985.

52. **Christ, M., Eisen, C., Aktas, J., Theisen, K., Wehling, M.,** The inositol-1,4,5-trisphosphate system is involved in rapid nongenomic effects of aldosterone in human mononuclear leukocytes. *J. Clin. Endocrinol. Metab.*, 77, 1452, 1993.

53. **Sadler, S. E., Maller, J. L.,** Progesterone inhibits adenylate cyclase in Xenopus oocytes, *J. Biol. Chem.*, 256, 63–68, 1981.

54. **Maller, J. L., Krebs, E. G.,** Progesterone-stimulated meiotic cell division in Xenopus oocytes. *J. Biol. Chem.*, 252: 1712, 1977.

55. **Steiner, A., Locher, R., Sachinidis, A., Vetter, W.,** Cortisol-stimulated phosphoinositide metabolism in vascular smooth muscle cells: a role for glucocorticoids in blood pressure control? *J. Hypertension*, 7, S140, 1989.

56. **Gelfand, E. W., Mills, G. B., Cheung, R. K., Lee, J. W., Grinstein, S.,** Transmembrane ion fluxes during activation of human T lymphocytes: role of Ca^{2+}, Na^+/H^+ exchange and phospholipid turnover, *Immunol. Rev.*, 95, 59, 1987.

57. **Tsien, R. Y., Pozzan, T., Rink, T. Y.,** T-cell mitogens cause early changes in cytoplasmic free Ca^{2+} and membrane potential in lymphocytes, *Nature,* 295, 68, 1982.

58. **Berk, B. C., Brock, T. A., Gimbrone, M. A., Alexander, R. W.,** Early agonist-mediated ionic events in cultured vascular smooth muscle cells, *J. Biol. Chem.,* 262, 5065, 1987.

59. **Grupp, I., Im, W.-B., Lee, C. O., Lee, S. W., Pecker, M. S., Schwartz, A.,** Relation of sodium pump inhibition to positive inotropy at low concentrations of ouabain in rat heart muscle, *J. Physiol.,* 360, 149, 1985.

60. **Deutsch, C., Price, M. A.,** Cell calcium in human peripheral blood lymphocytes and the effect of mitogen, *Biochim. Biophys. Acta,* 687, 211, 1982.

61. **Grinstein, S., Goetz, J. D.,** Control of free cytoplasmic calcium by intracellular pH in rat lymphocytes, *Biochim. Biophys. Acta,* 819, 267, 1985.

62. **Garty, H., Benos, D. J.,** Characteristics and regulatory mechanism of the amiloride-blockable Na$^+$ channel, *Am. Physiol. Soc.,* 68, 309, 1988.

63. **Sariban-Sohraby, S., Benos, D. J.,** The amiloride-sensitive sodium channel, *Am. J. Physiol.,* 250, C175, 1986.

64. **Eaton, D. C.,** Intracellular sodium ion activity and sodium transport in rabbit urinary bladder, *J. Physiol. Lond.,* 316, 527, 1981.

65. **Crabbé, J.,** Site of action of aldosterone on the bladder of the toad, *Nature,* 200, 787, 1963.

66. **Pietras, R. J., Szego, C. M.,** Specific binding sites for oestrogen at the outer surfaces of isolated endometrial cells, *Nature,* 253, 287, 1977.

67. **Kelly, M. J., Moss, R. L., Dudley, C. A., Fawcett, C. P.,** The specificity of the response of preoptic-septal area neurons to estrogen: 17α-estradiol versus 17β-estradiol and the response of extrahypothalamic neurons, *Brain Res.,* 30, 53, 1977.

68. **Dufy, B., Vincent, J.D., Fleury, H., Du-Pasquier, P., Gourdji, D., Tixier-Vidal, A.,** Membrane effects of thyrotropin-releasing hormone and estrogen shown by intracellular recording from pituitary cells, *Science,* 204, 509, 1979.

69. **Meiri, H.,** Acetylcholine release by progesterone in the brain, *Brain Res.,* 385, 193, 1986.

70. **Ramirez, V. D., Dluzen, D.,** Is progesterone a pre-hormone in the CNS? *J. Steroid. Biochem.,* 27, 589, 1987.

71. **Barsony, J., Marx, S. J.,** Rapid accumulation of cyclic GMP near activated vitamin D receptors, *Proc. Natl. Acad. Sci. U.S.A.,* 88, 1436, 1991.

72. **Petzel, D., Ganz, M. B., Nestler, E. J., Lewis, J. J., Goldenring, J., Akcicek, F., Hayslett, J. P.,** Correlates of aldosterone-induced increases in Ca$_i^{2+}$ and I$_{sc}$ suggest that Ca$_i^{2+}$ is the second messenger for stimulation of apical membrane conductance, *J. Clin. Invest.,* 89, 150, 1992.

73. **Morley, P., Whitfield, J. F., Vanderhyden, B. C., Tsang, B. K., Schwartz, J. L.,** A new, nongenomic estrogen action — the rapid release of intracellular calcium, *Endocrinology,* 131, 1305, 1992.

74. **Suyemitsu, T., Terayama, H.,** Specific binding sites for natural glucocorticoids in plasma membranes of rat liver, *Endocrinology,* 96, 1499, 1975.

75. **Schaeffer, J. M., Stevens, S., Smith, R. G., Hsueh, A. J.,** Binding of 2-hydroxyestradiol to rat anterior pituitary cell membranes, *J. Biol. Chem.,* 255, 9838, 1980.

76. **Blondeau, J. P., Baulieu, E. E.,** Progesterone receptor characterized by photoaffinity labeling in the plasma membrane of Xenopus laevis oocytes, *Biochem. J.,* 219, 785, 1984.

77. **Bression, D., Michard, M., Le Dafniet, M., Pagesy, P., Peillon, F.,** Evidence for a specific estradiol binding site on rat pituitary membranes, *Endocrinology,* 119, 1048, 1986.

78. **Quelle, F. W., Smith, R. V., Hrycyna, C. A., Kaliban, T. D., Crooks, J. A., O'Brien, J. M.,** [^{3}H]-dexamethasone binding to plasma membrane-enriched fractions from liver of nonadrenalectomized rats, *Endocrinology,* 123, 1642, 1988.

79. **Ke, F. C., Ramirez, V. D.,** Binding of progesterone to nerve cell membranes of rat brain using progesterone conjugated to 125I-bovine serum albumin as a ligand, *J. Neurochem.,* 54, 467, 1990.

80. **Ibarrola, I., Alejandro, A., Marino, A., Sancho, M.J., Macarulla, J.M., Trueba, M.,** Characterization by photoaffinity labeling of a steroid binding protein in rat liver plasma membrane, *J. Membr. Biol.,* 125, 185, 1992.

81. **Wehling, M., Kuhnle, U., Weber, P. C., Armanini, D.,** Lack of effect of aldosterone on intracellular sodium and potassium in mononuclear leukocytes from patients with pseudohypoaldosteronism, *Clin. Endocrinol.,* 28, 67, 1988.

82. **Wehling, M., Kuhnle, U., Keller, U., Weber, P. C., Armanini, D.,** Inheritance of mineralocorticoid effector abnormalities of human mononuclear leukocytes in families with pseudohypoaldosteronism, *Clin. Endocrinol.,* 31, 597, 1989.

83. **Armanini, D., Kuhnle, U., Strasser, T., Dorr, H., Butenandt, I., Weber, P. C., Stockigt, J. R., Pearce, P., Funder, J. W.,** Pseudohypoaldosteronism: demonstration of aldosterone receptor deficiency. *N. Engl. J. Med.,* 313, 1178, 1985.

84. **Armanini, D., Wehling, M., DaDalt, L., Zennaro, M., Scali, U., Keller, U., Pratesi, C., Mantero, F., Kuhnle, U.,** Pseudohypoaldosteronism and mineralocorticoid receptor abnormalities, *J. Steroid Biochem. Mol. Biol.,* 40, 363, 1991.

85. **Roy, C.,** Pseudohypoaldosteronisme familial, *Arch. Franç. Pédiatrie,* 34, 37, 1977.

86. **Limal, J. M., Rappaport, R., Dechaux, M., Riffaud, C., Morin, C.,** Familial dominant pseudohypoaldosteronism, *Lancet,* I, 51, 1978.

87. **Hanukoglu, A., Fried, D., Gotlieb, A.,** Inheritance of pseudohypoaldosteronism, *Lancet,* I, 1359, 1978.

88. **Armanini, D., Witzgall, H., Wehling, M., Kuhnle, U., Weber, P. C.,** Aldosterone receptors in different types of primary hyperaldosteronism, *J. Clin. Endocrinol. Metab.,* 65, 101, 1987.

89. **Knox, W. H., Sen, A. K.,** Mechanism of action of aldosterone with particular reference to Na-K-ATPase, *Annu. N.Y. Acad. Sci.,* 242, 471, 1974.

90. **Wehling, M., Kuhls, S., Kuhnle, U., Theisen, K.,** Effects of aldosterone on intralymphocytic sodium and potassium in patients with essential hypertension, *Klin. Wochenschr.,* 68, 71, 1990.

91. **Edmondson, R. P. S., Hilton, P. J., Thomas, R. D., Patrick, J,. Jones, N. F.,** Abnormal leucocyte composition and sodium transport in essential hypertension, *Lancet,* I, 1003, 1975.

92. **Ambrosioni, E., Costa, F. V., Montebugnoli, L., Tartagni, F., Magnani, B.,** Increased intralymphocytic sodium content in essential hypertension: an index of impaired Na cellular metabolism, *Clin. Sci.,* 61, 181, 1981.

93. **Schmouder, R. L., Weder, A. B.,** Platelet sodium-proton exchange is increased in essential hypertension, *J. Hypertension,* 7, 325, 1989.

94. **Feig, P. U., D'Occhio, A., Boylan, J. W.,** Lymphocyte membrane sodium-proton exchange in spontaneously hypertensive rats, *Hypertension,* 9, 282, 1987.

95. **Wehling, M., Käsmayr, J., Theisen, K.,** Is the sodium-proton-exchanger stimulated in lymphocytes from patients with essential hypertension? *J. Hypertension,* 9, 519, 1991.

96. **Edwards, C. R. W.,** Renal 11β-OH-steroiddehydrogenase: a mechanism ensuring mineralocorticoid specificity, *Hormone Res.,* 34, 3, 1990.

97. **Wehling, M., Christ, M., Gerzer, R.,** Aldosterone-specific membrane receptors and related rapid, non-genomic effects, *Trends Pharmacol. Sci.,* 14, 1, 1993.

INDEX

U

Urinary acid excretion, 31
Urinary acidification, 31, 66–67
Ussing's chambers, 23, 26

V

Vascular smooth muscle cells
    aldosterone, 112–114, 144
    inositol-1,4,5-triphosphate, 132, 137
Vasopressin, 58
Ventral tegmental area, steroid hormone
    activity, 91–92

Ventromedial hypothalamus, 90–91
VSMC
    aldosterone, 112–114, 144
    inositol-1,4,5-triphosphate, 132, 137

X

Xenopus studies, 33, 35, 55–56, 88, 133

Z

Zinc finger structures, 20